ARTIST'S CONCEPTION OF THE MID-PACIFIC MOUNTAINS

If the Pacific Ocean were drained away, the mile-deep sunken islands would emerge as truncated volcanic cones. The original oil painting is by the distinguished scientific illustrator Chesley Bonestell and is based on part of the bathymetric chart of the Mid-Pacific range (Pl. 11).

The Geological Society of America
Memoir 64

SUNKEN ISLANDS OF THE MID-PACIFIC MOUNTAINS

BY
EDWIN L. HAMILTON
U. S. Navy Electronics Laboratory, San Diego, Calif.

March 10, 1956

Made in the United States of America

PRINTED BY WAVERLY PRESS, INC.
BALTIMORE, MD.

PUBLISHED BY THE GEOLOGICAL SOCIETY OF AMERICA
Address all communications to The Geological Society of America
419 West 117 Street, New York 27, N. Y.

*The Memoir Series
of
The Geological Society of America
is made possible
through the bequest of
Richard Alexander Fullerton Penrose, Jr.*

PREFACE

The last great frontiers of physical geography and geology are the vast, almost unknown floors of the oceans. Although man has always feared and wondered about the mysterious depths, it was not until HMS CHALLENGER made its historic voyage from 1872 to 1876 that the foundations were laid for the sciences of Oceanography and Marine Geology. Marine Geology gained slow momentum during succeeding years, but today it is probably the fastest expanding field of geology.

One of the centers of submarine geologic investigations on the West Coast is in the San Diego area. Submarine geologists of the Scripps Institution of Oceanography, under Dr. Roger R. Revelle and Dr. Francis P. Shepard, and of the Sea Floor Studies Section of the U. S. Navy Electronics Laboratory, led by Dr. Robert S. Dietz, co-operate closely in sea-floor exploration.

Dr. Dietz interested the writer in the problem of the flat-topped seamounts of the Pacific and arranged for his participation in the Scripps Institution of Oceanography-U. S. Navy Electronics Laboratory Expedition to the Mid-Pacific (1950) which, as part of its program, explored five seamounts and a great submarine ridge in the north central Pacific. This expedition was the first of a series of deep-sea expeditions planned by Scripps Institution of Oceanography to explore the largely unknown waters and deep-sea bottom of the Pacific Ocean. The U. S. Navy Electronics Laboratory is co-operating in these explorations by participating in the initial planning and by sending personnel and equipment.

The Mid-Pacific Expedition of 1950 (short name: Mid-Pac) was a joint venture led by Dr. Revelle primarily to collect geologic and oceanographic information. One objective was to explore the bottom of the Pacific along the ships' tracks. This involved bottom sampling, recording of bottom profiles, and the survey and sampling of seamounts with special emphasis on flat-topped seamounts.

Prior to the start of the expedition a detailed study was made of the Oceanic Sounding Sheets of the Navy Hydrographic Office which contain a wealth of new soundings taken by the Navy during World War II. These new soundings have never been adequately evaluated and published; they therefore furnish a fruitful field of bathymetric research. Their study and study of published charts indicated that an area about 1000 nautical miles west of Hawaii contains a great submarine mountain range surmounted by flat-topped seamounts. Exploration of seamounts was therefore concentrated in this area.

The Scripps-Navy Expedition left San Diego in late July, 1950, and arrived in Pearl Harbor, Oahu, a month later after extensive oceanographic, meteorological, bottom-profile, and sediment investigations which took the expedition down to 4° N. Lat. The U.S.S. EPCE (R) 857 and the Scripps Research Vessel HORIZON began the voyage, but the U.S.S. EPCE (R) 857 suffered breakdown on the first leg of the trip and was left in Pearl Harbor for repairs. The HORIZON, with all of the scientific party aboard, left port for single-ship exploration. Plate 10 shows the general area of investigation west of Hawaii. The area of the five guyots is shown at large scale in Plate 11. After completion of seismic work at Bikini Atoll in September the expedition broke up and most of the scientific party flew back to the West Coast from Kwajalein.

This report is concerned with the data and samples collected on the flat-topped seamounts and with the proof that these seamounts are an ancient chain of islands now sunk a mile deep in the Middle Pacific.

This report is a contribution from the Scripps Institution of Oceanography, New Series No. 830. U. S. Navy Electronics Laboratory Professional Contribution No. 7.

ACKNOWLEDGMENTS

The writer gratefully acknowledges the assistance and advice of the following persons:

Dr. Robert S. Dietz, who interested the writer in the problem of the guyots and who acted as technical advisor; Dr. Siemon W. Muller of Stanford University who acted as advisor on the doctoral dissertation of which this paper is the major portion; the scientific party and crew of the R/V Horizon; Dr. Roger Revelle who showed outstanding leadership as Director of the expedition.

At the U. S. Navy Electronics Laboratory, San Diego, where the general facilities were generously made available to him, the writer is grateful for many long discussions and advice from Dr. R. S. Dietz and Dr. H. W. Menard; for drafting assistance from Mr. G. L. Prible; for typing assistance from Mrs. Anna L. Moore and Mrs. Helen M. Steward; and for photographs taken by Mr. G. S. Anderson and Mr. J. H. Sneed of the Navy Electronics Laboratory Photographic Section.

The writer further acknowledges the assistance of Dr. A. Myra Keen who advised on the fossil fauna with Dr. Muller; of Mr. H. A. Hubbard who photographed some of the fossils; of Dr. John W. Wells of Cornell University who checked the writer's identifications of the reef coral and made several identifications, and of his colleagues at the University of Washington who aided by their helpful discussions and advice.

The writer appreciates the critical reading of the manuscript by Drs. R. S. Dietz, H. W. Menard, M. N. Bramlette, F. B Phleger, F. P. Shepard, R. R. Revelle, J. W. Wells, H. H. Hess, S. W. Muller, A. M. Keen, Miss F. L. Parker, Mr. E. C. Buffington, Mr. R. F. Dill, Mr. A. J. Carsola, and Mr. George Shumway.

CONTENTS

	Page
ABSTRACT	1
INTRODUCTION	1
General statement	1
Previous investigations	2
PHYSIOGRAPHY OF THE MID-PACIFIC MOUNTAINS	4
GEOLOGY OF THE GUYOTS	5
Introduction	5
Horizon Guyot	5
Guyot 19171	9
Guyot 20171	11
Cores at MP 27	12
Hess Guyot	14
Cape Johnson Guyot	18
PALEONTOLOGY	22
Introduction	22
Megafossils	22
General	22
Coral	23
Rudistids	26
Stromatoporoids	27
Gastropods	28
Echinoid	28
Miscellaneous	28
Conclusions	29
Microfossils	29
General discussion	29
Age, affinities, and ecology	30
LITHOLOGY AND SEDIMENTS	31
General	31
Igneous rocks	32
Limestone and calcareous oozes	33
General	33
Globigerina ooze	33
Detrital limestone	35
Silicified limestone	35
Phosphatized limestone	35
MANGANESE DIOXIDE	36
GEOMORPHOLOGY	38
Introduction	38
Geomorphology of the guyots	39
Truncation of the guyots	41
Geomorphology of the Mid-Pacific Mountains	43
CAUSES OF SUBMERGENCE	44
BEARING OF CONCLUSIONS ON CORAL-ATOLL HYPOTHESES	48
MODIFICATION OF THEORIES OF FAUNAL MIGRATION	50
SUMMARY AND CONCLUSIONS	53
GEOLOGIC HISTORY OF THE MID-PACIFIC GUYOTS	54
APPENDIX A. SYSTEMATIC PALEONTOLOGY	57
Part 1. Megafossils	57
Part 2. Faunal lists of Foraminifera	68

x   SUNKEN ISLANDS OF THE MID-PACIFIC MOUNTAINS

Appendix B. Lithology and Sediments... 75
Appendix C. Sampling Equipment and Techniques.................................. 81
Appendix D. Echo Sounding... 83
Appendix E. Location at Sea... 85
References Cited... 87
Index.. 93

ILLUSTRATIONS

Plates

Plate
1.—Artist's conception of the Mid-Pacific Mountains..........................Frontispiece
                                                                            Following page
2.—Dredge hauls from the flat-topped seamounts...⎫
3.—Dredge hauls from the flat-topped seamounts...⎬ 70
4.—Rock types from the dredge hauls...⎭
                                                                            Facing page
5.—Fossils from dredge hauls... 71
6.—Fossils from dredge hauls... 72
7.—Fossils from dredge hauls... 73
8.—Fossils from dredge hauls... 74
9.—Fossils from dredge hauls... 75
10.—Bathymetric chart between Hawaiian and Marshall islands................⎫
11.—Bathymetric chart of a part of the Mid-Pacific Mountains..................⎬ In pocket,
12.—Bathymetric chart of the northwest part of the Mid-Pacific Mountains........⎨ back of book
13.—Bathymetric chart of the southwest part of the Mid-Pacific Mountains........⎭

Figures

Figure   Page
1.—Profile of a typical central Pacific guyot... 2
2.—Areal distribution of supposed guyots in the Pacific................................ 3
3.—Profiles of Horizon Guyot.. 7
4.—Profiles of Horizon Guyot.. 8
5.—Plan and profiles of Guyot 19171... 10
6.—Plan and profile of Guyot 20171.. 11
7.—Plan and profiles of Hess Guyot... 15
8.—Plan and profiles of Cape Johnson Guyot.. 19
9.—Cumulative frequency curves for bottom of core MP 27-2......................... 78
10.—Cumulative frequency curve for top of core MP 27-2............................. 79
11.—Cumulative frequency curve from analysis of sandstone from MP 37-A........ 79
12.—Curves for the correction of bottom slopes determined by echo sounder........ 84

Tables

Table   Page
1.—Data on sampling stations 25–27.. 6
2.—Data on sampling stations 27–33... 13
3.—Data on sampling stations 33–37... 16
4.—Data on sampling stations 37–38... 17
5.—Range chart of the Cretaceous fauna of the Mid-Pac Mountains guyots......... 21

ABSTRACT

One of the objectives of the 1950 expedition to Bikini carried on by the Scripps Institution of Oceanography and the U. S. Navy Electronics Laboratory was to investigate previously located flat-topped seamounts (*guyots* of Hess, 1946) in an area 600–1100 miles west of Hawaii. Five of these seamounts were surveyed by echo sounder, dredged, and cored, and were found to be peaks on a great submarine range—the Mid-Pacific Mountains—which extends from Necker Island in the Hawaiian Islands to near Wake Island.

The flat-topped guyots are submerged to between 700 and 900 fathoms. The sides are concave upward with slopes about 20° near the tops. The profiles are symmetrical with breaks in slope at 720–1150 fathoms. Rounded sand grains, pebbles, cobbles, and boulders of olivine basalt were dredged from the tops and across the breaks in slope. Sandstone and reef coral were dredged together at one station. Dredge hauls across the breaks in slope on the tops of two guyots brought up an integrated Cretaceous (Aptian-Cenomanian) fauna of reef coral, rudistids, stromatoporoids, gastropods, pelecypods, and an echinoid; these have affinities with the faunas of the Tethyan Province around the Gulf Coast and adjacent regions. In a core taken in a basin near one guyot, the basaltic gravel layers contained Upper Cretaceous (Campanian-Maestrichtian) Foraminifera. Paleocene and Eocene Foraminifera occur on the flat tops of four guyots.

The evidence indicates that in Cretaceous time the guyots were a chain of basaltic islands. These islands were wave-eroded to relatively flat banks on which a reef coral-rudistid fauna found lodgment and grew into reefs on and among the erosional debris. They never became fully developed atolls. The guyots were submerged during the Cretaceous to below the zone of reef-coral growth; finally they sank to present depth. The submergence is thought to have been due to regional subsidence of the sea bottom resulting for the most part from isostatic adjustments and subcrustal forces. A minor part of the submergence can probably be attributed to increase in ocean volume, sedimentation, and the compaction of soft sediments.

In general the findings support Darwin's Subsidence Theory for the formation of atolls; they furnish evidence for a deep Cretaceous Pacific Ocean; and they refute the hypothesis of transoceanic "sunken continents" used to explain faunal migrations, at the same time suggesting the possibility of "island stepping stones".

INTRODUCTION

GENERAL STATEMENT

Prior to the 1950 Scripps Institution-U. S. Navy Expedition, information about the flat-topped seamounts was tantalizingly scarce. The morphology of many of the features, based on echo-sounder profiles, was about all the information available. All previous investigators agreed that these ocean-bottom features were probably wave-eroded sunken islands, but the conclusions lacked concrete proof, and the problem demanded further exploration. The study of erosional debris would confirm the evidence of the truncated tops as eroded surfaces and would shed light on the rock composition of the seamounts. The presence of fossils might determine an age of truncation or subsidence.

The origin and subsequent history of the flat-topped seamounts have far-reaching implications in geology and are related to questions on (1) the age, stability, and tectonics of ocean basins, (2) coral-reef formation, (3) changes in the volume and depth of the ocean during geologic time, (4) isostasy.

PREVIOUS INVESTIGATIONS

The flat-topped seamounts of the Pacific first attracted widespread scientific attention in 1946 when Harry H. Hess of Princeton University published his paper

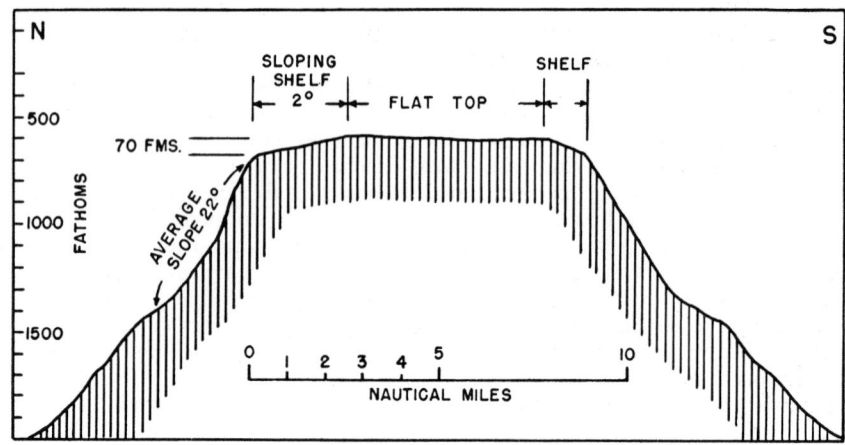

FIGURE 1.—*Profile of a typical central Pacific guyot*
(From Hess, 1946)

"Drowned Ancient Islands of the Pacific Basin". While on wartime cruises in the Pacific, Hess noted these anomalous flat-topped seamounts from echo-sounder traces. He actually crossed 20 of these features and later inferred the existence of more than 140 of them from the study of Navy Hydrographic Office Charts. The general characteristics he noted are shown in Figure 1. The distribution of the features in the Western Pacific (according to Hess) is shown in Figure 2. The flat-topped seamounts with which Hess was mainly concerned range from 520 to 960 fathoms below sea level. In brief, Hess believed that the seamounts were volcanic, bare of sediments, free of coral, and once had been islands which were eroded to flat tops by wave action. He regarded the flat-topped features as extremely ancient islands now submerged by rise of the sea because of sedimentation. He postulated a stable Pacific Basin and dated the islands as Precambrian because he believed that they must have been in a sea free of calcium carbonate-secreting organisms which, otherwise, would have kept the islands at the surface as sea level rose. Hess named the flat-topped seamounts "guyots" after the nineteenth century geographer, Arnold Guyot, and this name has become widely accepted.

Dr. Hess's intriguing paper created immediate and widespread interest. Some of the seamounts in the Gulf of Alaska described in Murray's (1941) purely descriptive paper were recognized by Hess as guyots. Murray failed to make known the fact that some of the seamounts had anomalous flat tops. Menard and Dietz (1951) have discussed 11 truncated seamounts in the Gulf of Alaska and postulate a pre-Middle Tertiary age for these guyots on the assumption that they are older than the Aleutian

Trench. The submergence of these Gulf of Alaska guyots is thought to have been due to rise of sea level, a general subsidence of the sea floor, or the local subsidence of individual seamounts, or a combination of these factors. Some guyots of the Gulf of Alaska are almost accordant at about 400 fathoms. The authors believe that lack

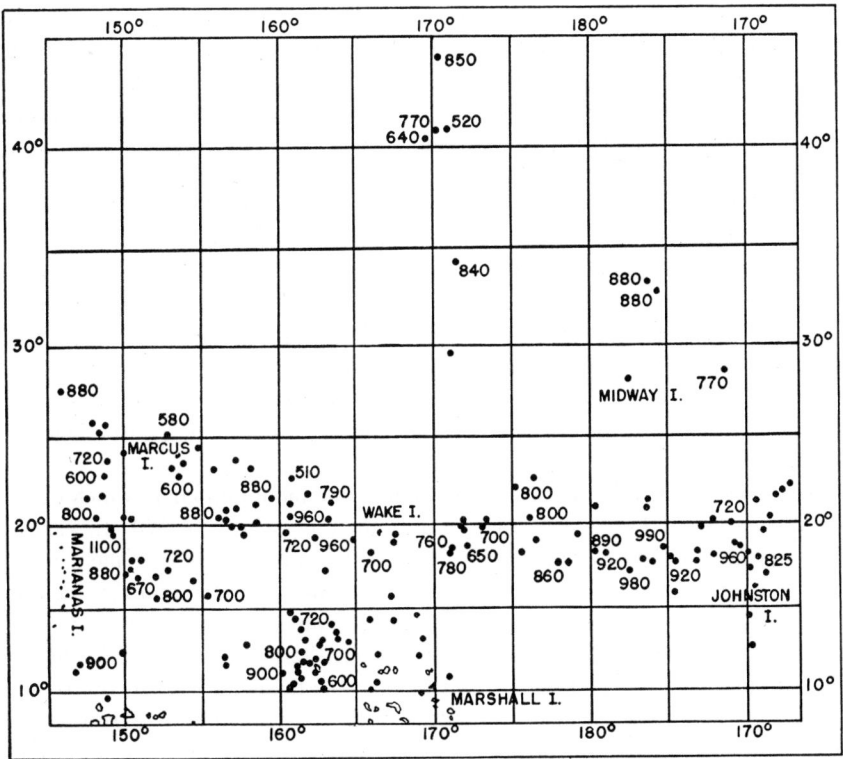

FIGURE 2.—*Areal distribution of supposed guyots in the western and central Pacific*
The numerals next to some of the guyots indicate the depth in fathoms to the flat upper surface (from Hess, 1946).

of accordance of some guyots may be due to truncation of a volcanic cone, the building up of new islands, and their subsequent truncation at a level higher than the already drowned features.

Carsola and Dietz (1952) have studied Erben and Fieberling guyots (800 and 600 miles, respectively, west of San Diego). These are truncated, basaltic volcanoes with flat tops at 400 and 280 fathoms. The drowned position of these wave-eroded surfaces was ascribed mainly to local isostatic subsidence.

Emery *et al.* (1954) have described the numerous flat-topped seamounts of the Northern Marshall Islands. They surveyed 14 guyots in the Northern Marshalls which have flat tops between 470 and 870 fathoms; 23 others have been incompletely surveyed. Because of morphology, Emery *et al.* believe them to be volcanic in origin. On the 1950 Mid-Pac expedition some basalt was dredged from one of these seamounts. Emery *et al.* postulated some regional subsidence and individual sinking of the volcanic masses to explain submergence.

PHYSIOGRAPHY OF THE MID-PACIFIC MOUNTAINS

The Scripps-Navy Expedition to the Mid-Pacific was the second cruise to gather bottom samples in the Mid-Pacific Mountains. The first was the U.S.S. TUSCARORA in 1874 (Belknap, 1874). The route of the TUSCARORA is indicated in Plate 10. The TUSCARORA expedition was interested in cable routes in the Western Pacific and made soundings and collected bottom samples. The sediments collected were globigerina ooze and red clay. Sediments listed as "white coral and sand" were obtained by the writer from the National Museum and proved to be globigerina ooze.

The Mid-Pac expedition was the first to explore the main-ridge area of the Mid-Pacific Mountains (Plate 11). A ridge or line of shallow seamounts is indicated on a bathymetric chart of the U. S. Navy (Bryan, 1940) and on the 1939 and 1950 charts of the International Hydrographic Bureau (Monaco). Hess (1946, p. 779) indicated guyots in the area arising from "hilly areas". The contribution of the Mid-Pacific Expedition, therefore, was not the "discovery" of a great undersea range, but the collection of proof that previously known shallow seamounts were in reality connected, not separated by deep passes, and the clearer definition of a great ridge upon which the seamounts rise as peaks.

The Mid-Pacific Mountains are a great submarine mountain chain which extends from Necker Island in the Hawaiian Islands to the vicinity of 170° E. Long., a distance of about 1500 nautical miles. This mountain chain is extremely narrow (about 25 nautical miles wide) near Necker Island. Between 180° and 175° E. Long. it reaches its maximum width at about 600 nautical miles. The mountains are shown in their entirety on Plate 10 and at larger scale on Plates 11, 12, and 13. The Mid-Pacific Mountains are not uniformly continuous; the main ridge is cut by passes as deep as 2200 and 2500 fathoms. (*See* Plate 10.) Flat-topped and sharply pointed peaks are present all along the main ridges. Summits of the peaks show no accordance. The Mid-Pacific Mountains are on the crest of a low, broad swell on the sea floor which is similar to the Hawaiian Swell (Dietz and Menard, 1953, p. 100). The lineation shown by the Mid-Pacific Mountains is generally east-west, not northwest-southeast which is the direction mentioned by many writers as the dominant orientation in the Pacific.

The part of the Mid-Pacific Mountains which is the subject of this paper (Pl. 11) starts in the east as two elongate ridges which merge into one long continuous ridge about 600 nautical miles long and 43–92 nautical miles wide with a relief of over 13,000 feet above the ocean floor to the north and south.

The Bathymetric Chart of the problem area (Pl. 11) was based on soundings of the two expedition ships and the soundings on the published charts and sheets of the U. S. Navy Hydrographic Office and the Coast and Geodetic Survey. The main ridge was located and defined by the R/V HORIZON in zig-zag crossings. Soundings on parts of the ridge permitted reliable contouring; on other parts where soundings were few, or entirely missing, "logical contouring" was used based on the main trends of the ridge.

The findings of Mid-Pac thus show for the first time that the Mid-Pacific Mountains form a major, well-defined geomorphic feature of the earth's crust—a great underwater mountain chain surmounted by sharp peaks and ridges and by anomalous flat-topped "guyots".

GEOLOGY OF THE GUYOTS
INTRODUCTION

Five guyots of the Mid-Pacific Mountains were explored by the Scripps-Navy Expedition (Plate 11). One (Horizon Guyot) appears to be a flat-topped ridge which merges to the west with the main ridge of the Mid-Pacific Mountains. The other four guyots are peaks on the main ridge. Each of the five guyots was surveyed by echo sounder by crossing and recrossing the flat surfaces. Samples were collected by dredging and coring. The stations occupied, equipment used, samples gathered, and other pertinent data are compiled in the tables. The appendices give systematic descriptions of fossils, detailed lithology, sampling equipment and techniques used, and a short résumé of echo sounding.

HORIZON GUYOT

Horizon Guyot is named after the Scripps Institution R/V HORIZON. From the bathymetric chart (Plate 11) it appears as a ridge which is aligned in a northeast-southwest direction. It is about 40 statute miles wide at its widest part, about 170 miles long, and is centered at 19°–20° N. Lat. and 169° W. Long. A saddle more than 1200 feet deep divides the ridge near its west end so that the western end appears as an oval seamount. Samples were collected on this western peak (Table 1). Farther to the west Horizon Guyot merges with the main ridge of the Mid-Pacific Mountains.

Horizon Guyot rises abruptly from the 2600-fathom-deep sea floor. It has a flat top with well-defined breaks in slope at about 985 fathoms on the main ridge and a break in slope on the west-end seamount at about 938 fathoms. The shallowest sounding is 774 fathoms on the eastern ridge.

Figures 3 and 4 show two profiles across the main ridge. (*See* Pl. 11.) Each profile is shown with no vertical exaggeration; a profile across the top is drawn with a vertical exaggeration of five times. These profiles show that Horizon Guyot has the same general characteristics, with the exception of shape, as the "typical guyot" (Fig. 1) of Hess. The steepest side slopes are near the top, just below the break in slope. In this portion the slopes are about 15°; below this slope the bottom flattens and merges with the deep-sea floor. The top is flat, but near the edge there appears to be the "sloping shelf" (Fig. 1) noted by Hess. This shelf has a slope of about 3°. Just above C' (Pl. 11) on the north side of the eastern ridge there is a steep slope of about 37°.

Eight sampling stations were occupied on the west end of Horizon Guyot. (*See* Table 1.) Coring operations were carried out at five of these stations; in each a core of globigerina ooze was taken. The longest core (31 inches) was collected at MP 25 E-1, just below the break in slope.

Dredge hauls were made in two places. The first (MP 25F-1) brought up a few, flat, highly altered volcanic rocks covered with manganese dioxide coatings up to 35 mm in thickness, together with a few manganese nodules. The center of one nodule

TABLE 1.—DATA ON SAMPLING STATIONS 25–27

MP station	Date	Position N. Lat. W. Long.	Depth (fathoms)	Sampler	Sample description	Field notes and remarks
25 A	9/1/50	19 02 169 42	850	Phleger corer	28.5-inch core globigerina ooze	On top of west end of Horizon Guyot
25 B	9/1/50	19 10 169 49	1140	Three snappers	Few grains of globigerina ooze and particles of MnO₂	Below break in slope; lost a dredge at this station
25 C	9/1/50	19 07 169 48	900	Emery-Dietz corer	6-inch core of globigerina ooze with a few pumice fragments	Just above break in slope
25 D	9/2/50	19 05 169 46	892	Emery-Dietz corer	Few grains of globigerina ooze and MnO₂ particles	Top of seamount; almost no sample
25 E-1	9/2/50	19 02 169 44	931	Phleger corer	31-inch core of globigerina ooze	Just above break in slope; Eocene foram fauna
25 E-2	9/2/50	19 02 169 44	950	Phleger corer	28-inch core globigerina ooze	Just above break in slope; Eocene foram fauna
25 F-1	9/2/50	19 07 169 44	950	Chain-bag dredge	Four flat, deeply altered volcanic rocks: average size 4" x 8" x 2"; small chips of manganese-coated rock; manganese nodules	Dredging across break in slope;
25 F-2	9/2/50	19 07 169 44	935–960	Chain-bag dredge	Manganese-coated, fresh and altered olivine basalt; fossil globigerina ooze; manganese nodules; silicified limestone; basalt pebbles	Dredging just below break in slope; thin-section example of 8 manganese nodule centers (subrounded olivine basalt and silicified limestone)
26 A-1	9/3/50	*	685	Phleger corer	Few grams of manganese chips and globigerina sand	On top of Guyot 19171; core barrel badly smashed
26 A-2	9/3/50	....	675–680	Chain-bag dredge	Three small pieces of MnO₂	Dredging on top of seamount
26 A-3	9/3/50	....	720–770	Chain-bag dredge	Manganese-coated olivine basalt boulders with rounded corners, manganese-coated fossil globigerina ooze rock; manganese nodules; silicified limestone; sandstone pebble; olivine basalt chips	Dredging across break in slope. Thin-section example of 7 manganese nodule centers: olivine basalt, silicified limestone; composite pebble of limestone and sandstone
26 B	9/3/50	....	705	Chain-bag dredge	Few manganese chips; small basalt pebble	Dredging up slope on top
27-1	9/4/50	....	2050	Phleger corer	23½-inch core: 0–2½ inch, globigerina ooze; 2½–12 inch, transition to red clay; 12–23½ inch, red clay	Coring in basin near Guyot 20171; many manganese flakes and minute nodules in core

* Positions restricted by U. S. Navy.

GEOLOGY OF THE GUYOTS

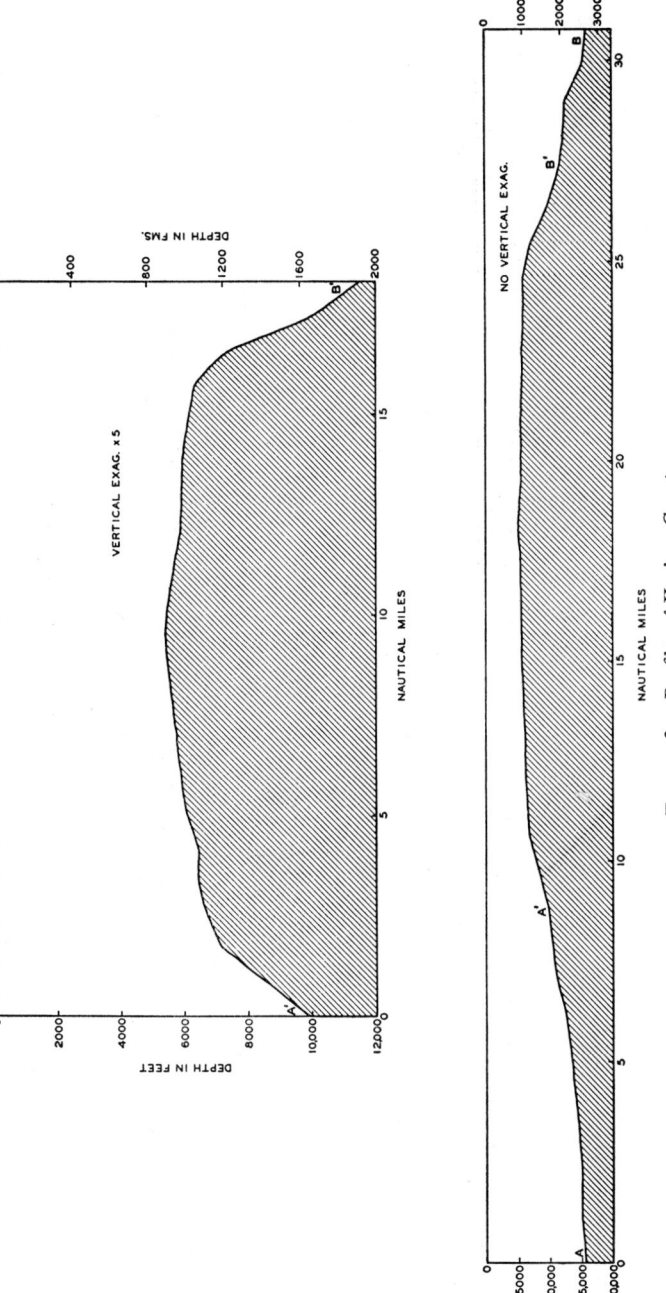

FIGURE 3.—*Profiles of Horizon Guyot*

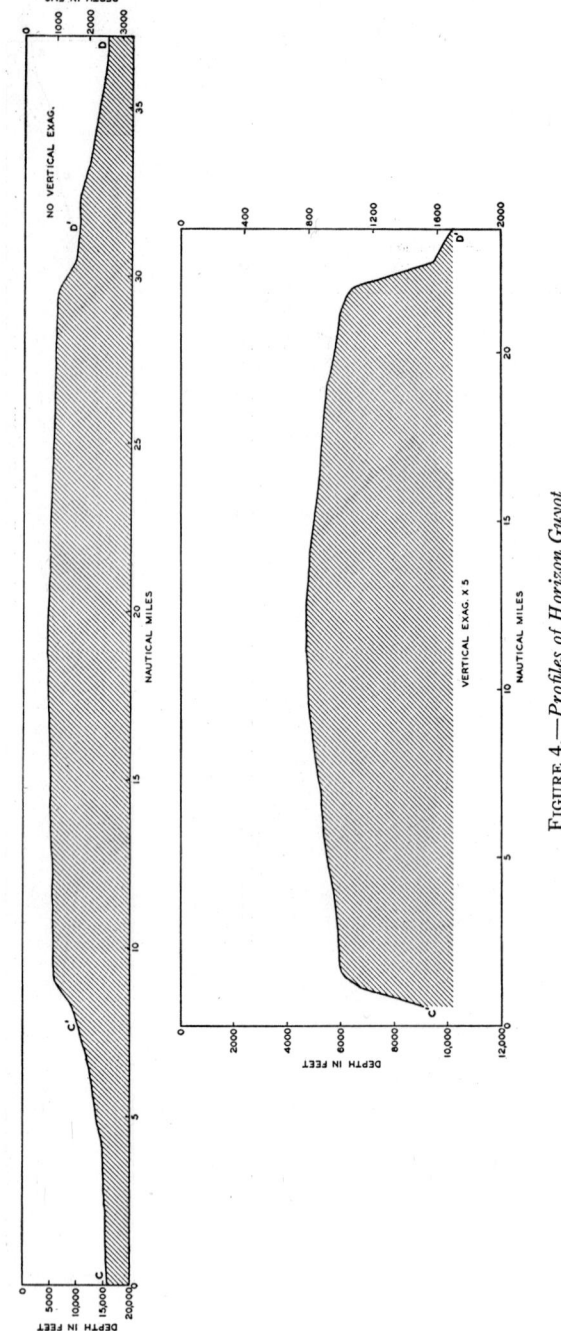

FIGURE 4.—*Profiles of Horizon Guyot*

was examined in thin section and was found to be a rounded pebble of partially phosphatized limestone.

The second dredge haul, just below the break in slope (MP 25F-2; Pl. 2, fig. 1) included fresh and altered olivine basalt, a boulder of broken volcanic rock cemented by calcareous ooze and red clay, and many manganese nodules. The centers of eight manganese nodules (examined in thin sections) proved to be subangular to subrounded olivine basalt pebbles or partially silicified or phosphatized limestone.

Several of the manganese-coated rocks contained cracks and pockets filled with globigerina ooze. This ooze was washed for Foraminifera. At MP 25 F-2 these cracks yielded a fauna ranging from Early Tertiary to Recent.

Two cores of particular interest were taken on a 2-degree slope just above the break in slope at the edge of the guyot. Each core contained the same pure Eocene fauna except in the top 1–2 inches which contained a mixture of Eocene species and modern tropical Pacific species. The present sea bottom at this locality on top of Horizon Guyot is, therefore, a mixture of Eocene and Recent species. One core (MP 25E-2) was 28 inches long and was a pure, unconsolidated Eocene globigerina ooze from the bottom to within 2 inches of the top. The second core (MP 25E-1), 31 inches long, was like the first except that the bottom 4 inches contained a mixture of Recent and Eocene species similar to the topmost portion. This could have been caused by the corer's hitting the bottom a second time, or the deposit cored may have been a slump. Both cores were taken at the same station but not necessarily at the same place because of drift of the ship on station. The Eocene fauna is listed in Appendix A, Part 2.

GUYOT 19171[1]

Guyot 19171 is a peak on the main ridge of the Mid-Pacific Mountains. It stands more than 10,000 feet above the deep-sea floor. The feature is flat-topped with breaks in slope between 720 and 749 fathoms. On the northwest side the slope near the top is about 14°. The shallowest sounding was 668 fathoms on top near the southwest side. The exact position of this guyot has been restricted by the Navy.

The bathymetric chart of Guyot 19171 (Fig. 5) shows the feature as an elongate seamount. It is about 35 nautical miles long and 20 nautical miles wide. This seamount is asymmetrical with the steeper slope to the northwest. Slopes on all sides are concave upward.

Figure 5 shows two profiles. The profile on the northeast side does not indicate a flat top. The shallowest sounding in this portion of the guyot was 766 fathoms. The second profile to the west shows a flattened surface whose shallowest sounding is 668 fathoms. It is possible that the eastern portion of the seamount was underwater while the higher western portion was being eroded. The surface of this guyot is not as flat as that of Hess and Cape Johnson guyots to the west. There is no evidence of the "sloping shelf" just above the break in slope which is found on some of the other guyots.

[1] A numbering system suggested by R. S. Dietz is used to identify unnamed seamounts. This method utilizes a number derived from the latitude and longitude of the feature. When the quadrant of the earth in which the seamount is located is not obvious, it can be designated by prefixing the number with NW for northwest, etc. By this method the location is defined to within a maximum limit of 43 nautical miles.

10 SUNKEN ISLANDS OF THE MID-PACIFIC MOUNTAINS

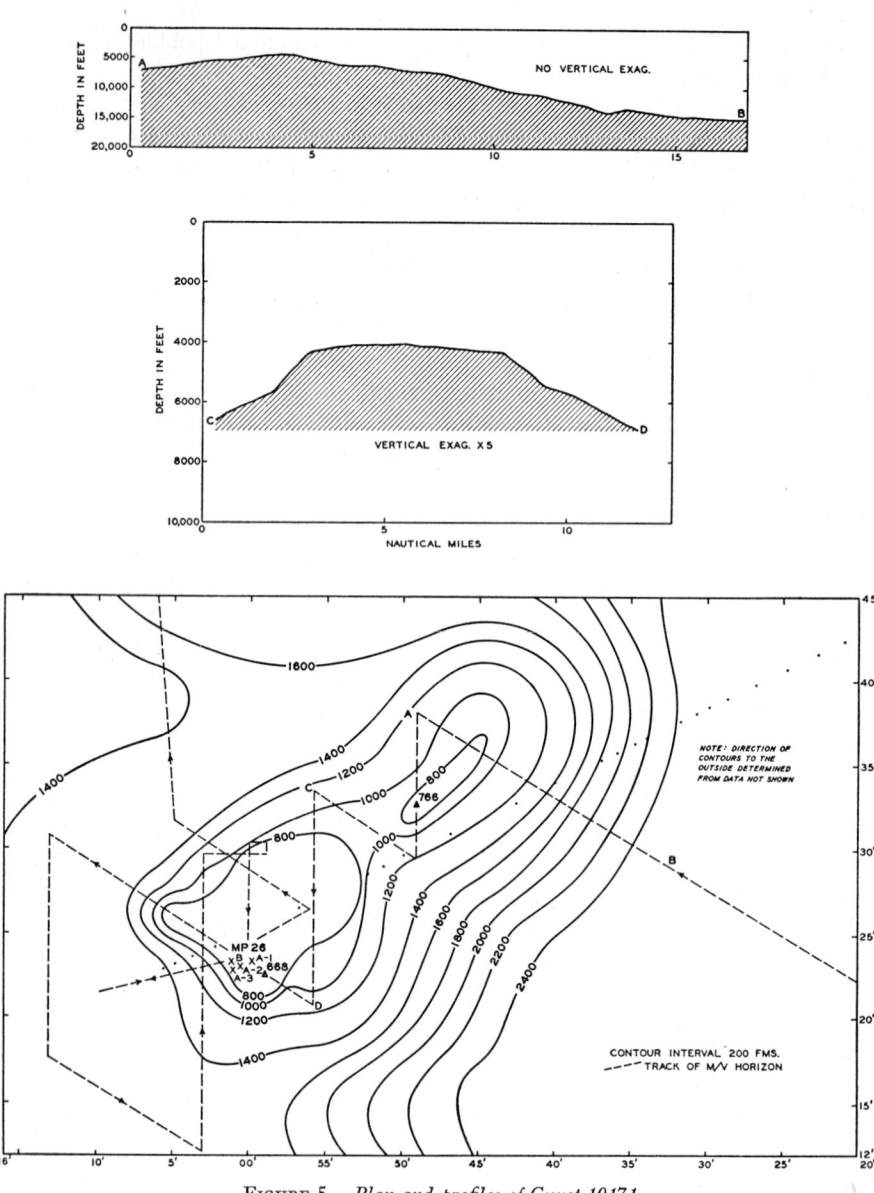

FIGURE 5.—*Plan and profiles of Guyot 19171*

Four sampling stations were occupied on Guyot 19171 (Fig. 5). The data and results from these stations are indicated in Table 1. One attempt was made to core with the Phleger Bottom Sampler. The core barrel was smashed and only a few manganese dioxide chips and a few grains of globigerina sand were collected.

Three dredge hauls were made. Two of them brought up small pieces of manganese dioxide and one basalt pebble. The third dredge haul (MP 26 A-3; Pl. 2, fig. 2) was very successful. Two large boulders of olivine basalt were brought up with thin coats of manganese dioxide (6-10 mm in thickness). The most interesting specimen was a large, manganese-covered boulder of indurated globigerina ooze. This haul also included many manganese nodules and slabs of olivine basalt.

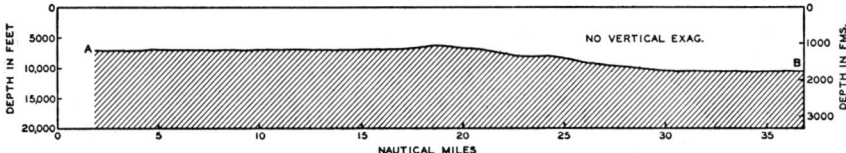

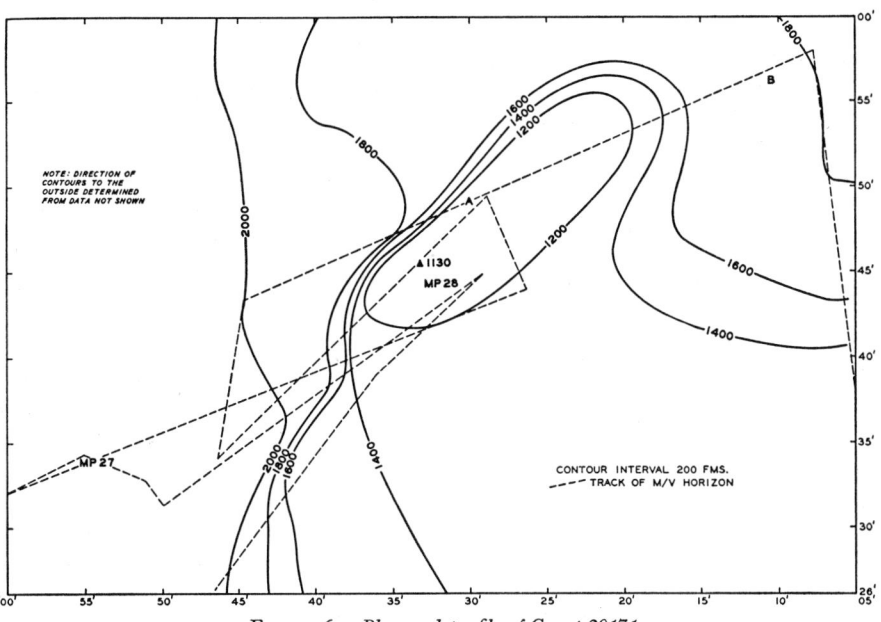

FIGURE 6.—*Plan and profile of Guyot 20171*

Seven manganese nodules were thin-sectioned, and these sections were examined petrographically. The centers of these nodules were mostly of hardened, partially phosphatized globigerina ooze. One center was a subangular pebble of olivine basalt and one was a composite pebble of limestone and sandstone. The sandstone was composed of detrital minerals from a basaltic source and rounded fragments of basalt.

The large indurated boulder of globigerina ooze contained a Lower Eocene foraminiferal fauna.

GUYOT 20171

Guyot 20171 is an elliptical, flat-topped peak on the main ridge of the Mid-Pacific Mountains (Fig. 6). The greatest relief (over 3600 feet) is on the northwest side.

Guyot 20171 is about 25 nautical miles long and about 7 nautical miles wide. It has a distinctly flat top a little deeper than 1100 fathoms. (*See* profile, Fig. 6.) There is a well-defined break in slope between 1120 and 1150 fathoms. The shallowest sounding was at 1075 fathoms. The steepness of side slopes near the top varies widely. The best-defined side slope (on the west end) was about 19°.

Guyot 20171 is the only guyot "discovered" by the Mid-Pacific Expedition in the region discussed and it is the most poorly defined one. Unfortunately the ship's track crossed the slopes at bad angles, and some unknown errors in navigation or plotting render an "objective" bathymetric chart impossible. The main features, however, probably approximate those shown in Figure 6. The exact position of this guyot has been restricted by the Navy.

One dredge haul (MP 28; Table 2) was made on top of the seamount near the west side. A few small manganese nodules and about a quart of globigerina ooze were the only sediments brought up. A study of thin sections of four manganese nodule centers revealed them to be highly altered, subrounded basalt pebbles. The globigerina ooze contained a foraminiferal fauna which was dominantly that of the modern tropical Pacific with rare specimens of *Globorotalia velascoensis*, *G. aragonensis*, and *G. crassata*, all Lower Tertiary planktonic forms.

CORES AT MP 27

Three cores were taken at MP 27. This station is in a basin or deep valley which is a re-entrant into the south side of the main ridge of the Mid-Pacific Mountains, below the slopes of Guyot 20171 at a depth of 2050 fathoms (Fig. 6; Pl. 11). The distance from MP 27 to the foot of the slope of Guyot 20171 is about 15 nautical miles. No soundings were available on three sides of MP 27, and consequently the topography surrounding the area is largely unknown. Undiscovered seamounts may be nearer MP 27 than is Guyot 20171.

The core at MP 27-1 was taken with the Phleger Bottom Sampler to test the bottom before sending down other equipment. A simplified description of this core is as follows:

Inches	
0 –½	cream-colored globigerina ooze with a few flakes and nodules of manganese dioxide
½– 2½	cream-colored globigerina ooze with many flakes and minute nodules of manganese dioxide
2½–12	transition of globigerina ooze to red clay
12 –23½	red clay

The next operation at this station was to core the bottom with a Kullenberg-type piston corer to which was attached a Phleger corer. The piston core was labelled MP 27-2 and the Phleger core MP 27-2P.

The bottom was successfully cored on this second operation, but the core extrusion device jammed in the core barrel during extrusion of MP 27-2 and made further extrusion impracticable. As a result MP 27-2 was carried on the fantail of the R/V HORIZON until the ship returned to San Diego. At the Scripps Institution Sedimentation Laboratory in November, 1950, it was found necessary to soften the core by wetting prior to its extrusion from the core barrel. The actual extrusion was difficult,

TABLE 2.—DATA ON SAMPLING STATIONS 27-33

MP station	Date	Position N. Lat. W. Long.	Depth (fathoms)	Sampler	Sample description	Field notes and remarks
27-2 & 2P	9/4/50	..........*	2050	Piston-type corer with Phleger corer attached	MP 27-2: Piston-type core of red clay with several layers of subrounded to angular, uncoated gravel from basaltic source: 27-2P: Phleger core with 0-7″ globigerina ooze; 7-34½″ red clay	Thin sections of nine pebbles from gravel in 27-2 analyzed as basalt; Phleger core with many flakes and minute nodules of manganese; coring in basin near Guyot 20171
28	9/5/50	....	1205-1170	Chain-bag dredge with pipe dredges	Few small manganese nodules; one quart of globigerina ooze	Dredging just below break in slope on Guyot 20171
29	9/6/50	18 04 173 11	2450	Phleger corer	7¼-inch core of red clay	Coring on south slopes of main ridge
30	9/6/50	18 27 173 14	2150	Phleger corer	25-inch core of red clay	Coring on south slopes of main ridge
31	9/7/50	18 31 173 17	1450	Phleger corer	8½-inch core of globigerina ooze	Coring on south slopes of main ridge
32	9/7/50	18 20 173 23	2110	Phleger corer	30-inch core of red clay	Coring on south slopes of main ridge
33 A	9/8/50	17 47.5 174 18	960-937	Chain-bag dredge with pipe dredges	Rock of rounded to subrounded calcareous fragments, foram tests, few small pelecypod shells, all cemented by CaCO₃; plates of "spongy" manganese and globigerina ooze	Dredging just below break in slope on Hess Guyot; calcareous rock resembles beach limestone found on the beaches of modern atolls.
33 B	9/8/50	17 47 174 20	940-925	Chain-bag dredge with pipe dredges	Three small rocks: encrusting stromatoporoids around limestone center; globigerina ooze	Dredging across break in slope toward center of Hess Guyot
33 C	9/8/50	17 49 174 17	910-920	Chain-bag dredge with pipe dredges	Rounded, white, soft rocks composed of foram tests, and finely divided CaCO₃; size range from sand to 9″ x 12″ x 5″; some had black manganese coating; coquina of shells of *Vermicularia* and CaCO₃; manganese nodules	Dredging near top center of Hess Guyot; thin sections of manganese nodules show centers of phosphatized limestone and olivine basalt.
33 D	9/9/50	17 54 174 16	980-956	Chain-bag dredge with pipe dredges	Coquina of fragments of *Vermicularia* (up to 2½ whorls) and finely divided calcareous ooze; manganese coatings up to 10 mm; globigerina ooze	Dredging below break in slope on Hess Guyot

* Positions restricted by U. S. Navy

and the top and bottom of the core were damaged. As a consequence of these difficulties the following description is only approximate:

Inches	
0 – 4	layer of gravel composed of rounded to subangular pebbles from a basaltic source held loosely together by red clay. (See Appendix B for mechanical analysis.)
4 –11	red clay with sand and small pebbles from basaltic source
11 –14	highly foraminiferal red clay
14 –22	red clay
22 –25	highly foraminiferal red clay
25 –28½	highly foraminiferal red clay with an increasing amount of sand and larger-sized rounded fragments of basaltic composition; pebble size increases downward
28½–30½	subangular to subrounded pebbles and sand from a basaltic source mixed with red clay
30½–34½	subrounded to subangular basaltic pebbles and cobbles mixed with red clay (Pl. 4, fig. 1)
34½–36½	basaltic gravel and sand mixed with red clay
36½–38	sand and small pebbles of basalt mixed with red clay

Complete mechanical analyses of the last four subdivisions are shown by cumulative curves (Fig. 9; Appendix B). The core apparently passed through, but was stopped by a layer of fresh-appearing subangular to subrounded gravel, the pebbles of which became larger downward, and penetrated into finer-sized material beneath. Thin sections were prepared of eight pebbles from the 30½–34½-inch division. These pebbles were of basalt and olivine basalt.

A description of MP 27-2P is included in Appendix B. In general this core was composed of 34½ inches of alternating layers of globigerina ooze and red clay with layers of sand and small pebbles near the bottom separated by small layers of globigerina ooze. No graded bedding was apparent.

Both MP 27-2 and MP 27-2P contain the same mixed Upper Cretaceous (Campanian-Maestrichtian) to Recent foraminiferal fauna from top to bottom. Several species each of *Globotruncana* and *Gümbelina* together with other species of the same age were the basis of this dating. (*See* "Paleontology, Microfossils.")

HESS GUYOT[2]

The center of Hess Guyot is at about 17°50′ N. Lat. and 174°15′ W. Long. (Fig. 7; Pl. 11; Tables 2, 3). Hess Guyot is a peak on the south side of the main ridge of the Mid-Pacific Mountains. It stands above the deep-sea floor to the south with a relief of about 9000 feet. The peak is conical with a distinctly flat top 8 nautical miles wide and 12 nautical miles long. The side slopes are concave upward and just below the break in slope they measure around 20°. On the south side the slopes gradually flatten and finally merge into the deep-sea floor. To the north the side slopes merge into the main ridge of the Mid-Pacific Mountains. There is a distinct break in slope on Hess Guyot around 930 fathoms. The shallowest sounding was 902 fathoms.

Three profiles of Hess Guyot are shown in Figure 7. The vertically exaggerated profile between B and C illustrates the remarkably flat top and distinct break in slope. The vertically exaggerated profile between E and F on the southeast side of the feature shows the flat top, breaks in slope, and two lower breaks in the concave upward side slope.

[2] Hess Guyot is named after Dr. Harry H. Hess of Princeton University, who discovered it while aboard the USS CAPE JOHNSON, in honor of his maintenance of scientific interest under difficult wartime conditions and of his paper (Hess, 1946) which first called attention to the guyots.

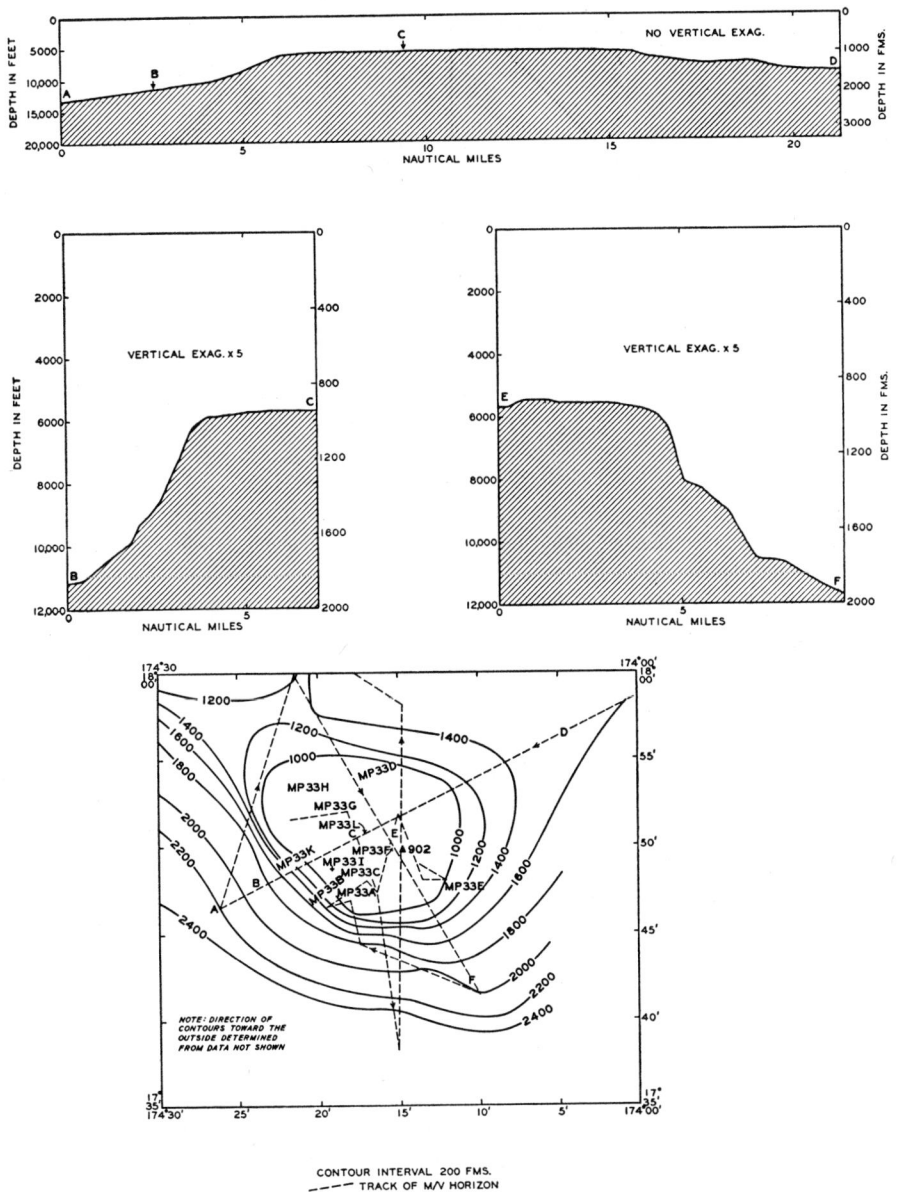

FIGURE 7.—*Plan and profiles of Hess Guyot*

Near MP 33 I on the southwest side of the top, a distinct bank was crossed which started at a depth of 930 fathoms on the inside, rose to a depth of 920 fathoms (or a height of 60 feet) and then dropped on the outside to a depth of 954 fathoms (a drop of 204 feet). This bank closely resembles the banks in the ship channel of modern atolls, where they are formed by growing coral and coral debris.

Twelve sampling stations were occupied on Hess Guyot. At six of these stations the top and side slopes near the top were cored with a Phleger Bottom Sampler. At

TABLE 3.—DATA ON SAMPLING STATIONS 33-37

MP station	Date	Position N. Lat. W. Long.	Depth (fathoms)	Sampler	Sample description	Field notes and remarks
33 E	9/9/50	17 48 174 12	1100	Phleger corer	4¾-inch core of globigerina ooze	Coring below break in slope on Hess Guyot
33 F	9/9/50	17 50 174 15	942	Phleger corer	15-inch core of globigerina ooze	Coring near center Hess Guyot
33 G	9/9/50	17 52 174 19	947	Phleger corer	38¼-inch core of globigerina ooze	Coring near center Hess Guyot
33 H	9/9/50	17 53 174 27	933	Phleger corer	21¼-inch core of globigerina ooze	Coring near west side Hess Guyot
33 I	9/9/50	17 48 174 19	954-920	Chain-bag dredge with pipe dredges	Two gallons of globigerina ooze; few gastropod shells with manganese coatings; 2 solitary living corals	Dredging across break in slope; Gastropods are *Vermicularia*
33 J	9/9/50	17 48 174 22	993	Phleger corer	No sample	
33 K			1250-990	Chain-bag dredge with pipe dredges	Fragments of fossil reef hexacorals; fossil echinoid; limestone with casts and molds of gastropods, pelecypods; coralline algae; most of the coral with manganese coatings up to 16 mm	Dredging up slope toward break in slope on southwest side Hess Guyot; 6 genera of Middle Cretaceous reef corals; echnoid (*Pyrina*) Jurassic-Eocene; 1 genus stromatoporoid
33 L	9/9/50	17 51 174 17	940	Phleger corer	17-inch core of globigerina ooze	Coring near center of Hess Guyot
34	9/10/50	18 38 175 03	567	Chain-bag dredge with pipe dredges	Fossil solitary coral	Dredging in sharp, uneroded peaks on north side range; this station had shallowest sounding
35-1	9/11/50	19 21 174 58	2700	Phleger corer	29½-inch core of red clay	Coring on flat bottom on north side of range
35-2	9/11/50	19 21 174 58	2650	Piston corer and Phleger corer	Piston core: 12 feet of red clay; Phleger core: red clay	Coring on flat bottom on north side of range
36	9/13/50	16 48 176 27	2700	Piston corer and Phleger corer	Piston core: 10 feet of red clay; Phleger core: red clay	Coring on flat bottom on south side of range
37 A	9/14/50	17 04 177 15	1100-1000	Chain-bag dredge with pipe dredges	Detrital and partially phosphatized fragments of reef coral; fossil solitary coral; fragments of gastropods and pelecypods; manganese-coated sandstone	Dredging up slope on Cape Johnson Guyot; 2 genera of Middle Cretaceous reef coral; sandstone from basaltic source

TABLE 4.—DATA ON SAMPLING STATIONS 37–38

MP station	Date	Position N. Lat. / W. Long.	Depth (fathoms)	Sampler	Sample description	Field notes and remarks
37 B	9/14/50	17 07 / 177 17	1030–980	Chain-bag dredge with pipe dredges	Two gallons of globigerina ooze and one cetacean earbone	Dredging toward break in slope on Cape Johnson Guyot
37 C	9/15/50	17 10 / 177 10	1120–1055	Chain-bag dredge with pipe dredges	About 200 lbs. of highly fossiliferous, partially phosphatized limestone; fossil calcareous ooze; manganese coatings up to 55 mm; many crusts flakes and nodules of manganese	Dredging toward break in slope; fossil fauna: one genus of M. Cretaceous reef coral; rudistids (Caprinidae); stromatoporoids; many fragments of gastropods and pelecypods
37 D	9/15/50	17 08 / 177 16	....	Chain-bag dredge	No sample	
37 E	9/15/50	17 08 / 177 16	982	Phleger corer	24-inch core of globigerina ooze	Coring just below break in slope on Cape Johnson Guyot
37 F	9/15/50	17 08 / 177 16	980	Phleger corer	26-inch core of globigerina ooze	Coring near center of Cape Johnson Guyot
37 G	9/15/50	17 07 / 177 18	1033	Phleger corer	10-inch core of globigerina ooze	Coring below break in slope on Cape Johnson Guyot
37 H	9/15/50	17 06 / 177 18	1170–1010	Chain-bag dredge	No sample	
37 I	9/15/50	17 05 / 177 13	991	Piston corer and Phleger corer	108½-inch core of globigerina ooze	Coring just below break in slope on Cape Johnson Guyot
38	9/16/50	19 02 / 177 18	2570	Piston corer	151½-inch red-clay core	Coring on flat bottom on north side of range

each of these six stations globigerina ooze was the only sediment cored. The longest core (38 inches) was taken near the center of the flat top.

Six dredge hauls were made with a chain bag dredge to which three small pipe dredges with canvas bags were attached. (*See* Appendix C.)

MP 33 A yielded fragments of detrital limestone composed of rounded and subrounded pebbles of limestone, tests of Foraminifera, and a few small pelecypod shells cemented together by calcite (Pl. 4, fig. 2).

At MP 33 B a small haul was made of three stromatoporoids (*Milleporidium*) encrusted around limestone centers.

At MP 33 C near the center of the guyot the dredge brought up large amounts of rounded, soft but indurated fossil globigerina ooze which ranged in size from pellets to boulders a foot in length (Pl. 2, fig. 3). When this haul came up over the side of the ship a stream of milky white water was pouring out of the dredge. When the dredge was placed on deck a wave broke over the fantail, passed through the dredge, and covered the deck with the milky white water. These fragments were apparently rounded in the dredge during the mile-long pull to the surface; the milky white water and scattered remnants of manganese dioxide crust indicated the ease with which the material was eroded.

MP 33 C also yielded a few manganese nodules—thin-section study revealed the centers to be of phosphatized limestone (Pl. 4, fig. 3c), and one well-rounded olivine basalt pebble (Pl. 4, fig. 3b). Also included in this haul was a coquina of gastropod fragments (*Vermicularia*) cemented together by calcareous ooze (Pl. 4, fig. 3a). The indurated globigerina ooze contained a well-preserved Paleocene fauna of fossil Foraminifera.

At MP 33 D a large amount of coquina of fragments of *Vermicularia* was dredged. The fragments (up to 2½ whorls) were cemented together by a finely divided calcareous ooze. Some of the material had coatings of manganese dioxide up to 10 mm thick.

MP 33 I yielded a few *Vermicularia* fragments coated with manganese dioxide and about two gallons of globigerina ooze from the small-pipe dredges.

At MP 33 K, from a depth of about 1000 fathoms just below the break in slope, many fragments of reef-building hexacorals were dredged (Pl. 3, fig. 1). These proved to be Middle Cretaceous (Aptian to Cenomanian). Most of the pieces were covered with coatings of manganese dioxide up to 16 mm in thickness and were not in the position of growth. This haul also yielded a well-preserved irregular echinoid of the genus *Pyrina* (Upper Jurassic to Paleocene), and aggregates of calcareous debris which contained casts and molds of pelecypods, echinoid spines, stromatoporoids, and coralline algae. (*See* Appendix A.)

CAPE JOHNSON GUYOT[3]

Cape Johnson Guyot is the only guyot of this paper to have been mentioned in the literature. Hess (1946, p. 782) gives a profile across Cape Johnson guyot and mentions that the "hummocky" surface might indicate sediments. The center of

---

[3] Cape Johnson Guyot is named after the USS CAPE JOHNSON, the ship on which Dr. Hess was navigator and commanding officer. This ship took the original line of soundings resulting in the discovery of Hess and Cape Johnson guyots.

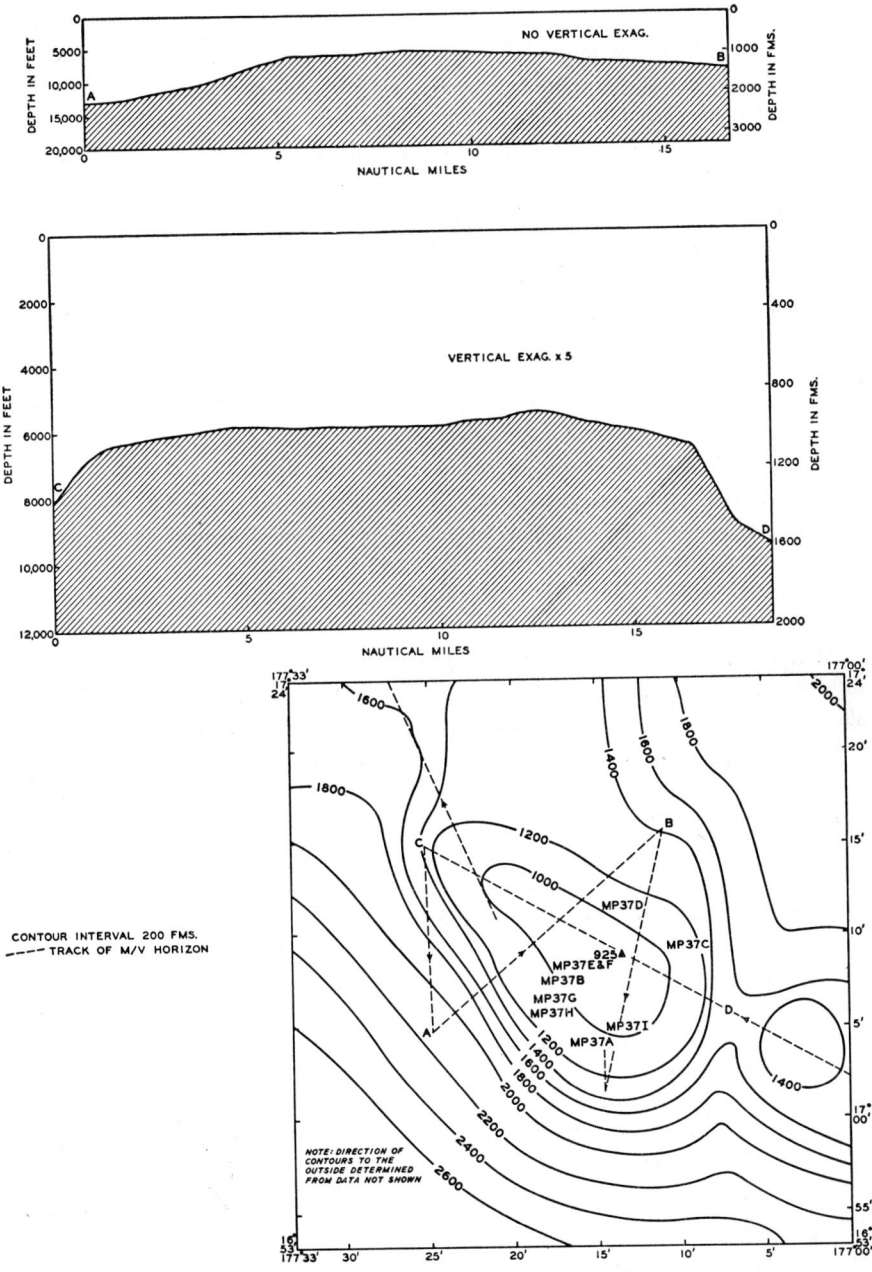

FIGURE 8.—*Plan and profiles of Cape Johnson Guyot*

this guyot is at about 17°08′ N. Lat. and 177°15′ W. Long. (Fig. 8; Pl. 11). Cape Johnson Guyot is a great peak on the south side of the main ridge of the Mid-Pacific Mountains. Its total relief with reference to the deep-sea floor to the south is more than 10,000 feet.

The peak is oval in plan, and within the 1000-fathom line the flat top measures about 6 by 12 nautical miles. The south and southwest sides are concave upward. Near the top the side slopes measure about 13°–16°; on the south and southwest sides of the guyot the slopes flatten and merge with the deep-sea floor below 2600 fathoms. To the north and east the side slopes merge into the main ridge of the Mid-Pacific Mountains. There is a distinct break in slope on Cape Johnson Guyot around 974 fathoms. The shallowest sounding was 925 fathoms near the center of the flat top.

The vertically exaggerated profile between C and D of Figure 8 illustrates the flat top and distinct breaks in slope. This profile shows a bank-line configuration on the southeast side of the guyot. This "bank" is 180 to over 300 feet higher than other parts of the top along the same profile. Such a bank is similar to those found on modern atolls where the coral and coral debris have grown and piled up near the edges of the top of the atoll. Profile A to B does not show this bank. Another bank was noticed on the east side of the top between MP 37 D and E, about 100 feet high.

Nine sampling stations were occupied on Cape Johnson Guyot. At three stations the guyot sediments were cored with the Phleger Bottom Sampler. All these cores were of globigerina ooze; the longest (26 inches) was taken near the center of the guyot. One Kullenberg type piston corer was used just below the break in slope on the southwest side at MP 37 I. This core was of pure globigerina ooze and measured 108½ inches.

At five stations the guyot was dredged with a chain-bag dredge with small-pipe dredges attached. At MP 37 A on the southwest side and just below the break in slope a haul (Pl. 3, fig. 2) was made which consisted of partially phosphatized fossil reef-building hexacorals covered with thin crusts of manganese; a fossil gastropod; a fossil solitary coral and a large manganese dioxide-coated sandstone rock.

The reef coral was detrital and not in the position of growth. In one rock several coral fragments and a gastropod were cemented together and covered with manganese dioxide. The coral had apparently rolled down slope from growth positions above.

The sandstone (Pl. 4, fig. 4) is composed of angular to subrounded detrital minerals from a basaltic source and small basaltic rock fragments. Almost all the grains are surrounded by a thin coat of $CaCO_3$; clay is also a cementing material. Mechanical analysis (Appendix B) of this sandstone reveals that it is well sorted and is skewed slightly to the coarse side. Slightly over 50 per cent of the grains have diameters between 0.125 and 0.500 mm. In the hand specimen there is a layered appearance owing to clay partings and some linear layers of manganese dioxide pellets.

The dredge haul at MP 37 B brought up one small earbone of a cetacean (Pl. 9, fig. 3) and about two gallons of globigerina ooze.

At station MP 37 C (just below the break in slope at about 1100 fathoms) the dredge brought up around 200 pounds of fossiliferous rock material (Pl. 3, fig. 3). Most of the rocks were coated with manganese dioxide up to 55 mm in thickness. Recognized in this haul were: indurated calcareous and globigerina ooze; phosphatized limestone; calcite fossils and fossils partially or wholly phosphatized; and crusts, flakes, and nodules of manganese dioxide. Many rocks contained all these elements cemented together by phosphatized and unphosphatized calcareous material and clay. Thin sections of some of the phosphatized limestone showed the outlines of

TABLE 5.—*Range chart of the Cretaceous fauna of the Mid-Pacific Mountains guyots*

SER	STAGES	MEGAFOSSILS	MP27-2 FORAMINIFERA (Part)
UPPER CRETACEOUS	DANIAN		
	MAESTRICHTIAN		
	SENONIAN CAMPANIAN		1, 12, 8, 2, 4, 9
	SANTONIAN		
	CONIACIAN		11
	TURONIAN		5, 3, 6, 7, 10, 13
	CENOMANIAN		1. GLOBOROTALIA VELASCOENSIS / 2. GLOBOTRUNCANA ARCA / 3. CANALICULATA / 4. CALCARATA / 5. FORNICATA / 6. MARGINATA / 7. VENTRICOSA / 8. GUMBELINA COSTULATA / 9. PLUMMERAE / 10. GLOBULOSA / 11. STRIATA / 12. ULTIMATUMIDA / 13. VENTILABRELLA AUSTINANA
	ALBIAN	ASTROCOENIA, MONTASTREA, MICROSOLENA, CYATHOPHORA, DIPLOASTREA, LOPHOSMILIA, BRACHYSERIS, NERINEA, CERITHIUM, TROCHUS, CAPRINA, CARDITA, PYRINA, MILLERORIDIUM, STROMOTOPORA, ACTINOSTROMA	
LOWER CRETACEOUS	APTIAN		
	BARREMIAN		
	NEOCOMIAN HAUTERIVIAN		
	VALANGINIAN		
	BERRIASIAN		

Foraminifera which confirmed the belief that the material was phosphatized globigerina ooze.

The other two dredge hauls obtained no samples.

Cape Johnson Guyot yielded by far the most fossil material. The fauna at MP 37 A consisted of two reef-building hexacoral genera (*Astrocoenia* and *Montastrea*), one solitary coral (*Lophosmilia*), and an extinct genus of gastropod (*Nerinea*).

The highly fossiliferous material of MP 37 C yielded an extinct genus of reef coral (*Brachyseris*), stromatoporoids (*Actinostroma*), gastropod and pelecypod shells, and abundant fragments and whole shells of rudistids of the Family Caprinidae.

The fossil material is Middle Cretaceous (Aptian to Cenomanian). Lower Tertiary to Recent planktonic species of Foraminifera were rare in the cores and in cracks inside the manganese-coated rocks.

PALEONTOLOGY

INTRODUCTION

This section includes discussions of the fossils dredged and cored on the guyots of the Mid-Pacific Mountains, their age, affinities, paleoecology, and the assemblages present. The basis of the overall age dating is shown in tabular form on the range chart (Table 5). The systematic descriptions of the fossils will be found in Appendix A.

MEGAFOSSILS

General.—All of the megafossils were dredged from Hess and Cape Johnson guyots.

Assemblages from Hess Guyot

MP Stat.	Locations*	Depth†	Genera	Class
33 B	Across	935	*Milleporidium*	Stromatoporoidea
33 C	Above (near top)	915	*Vermicularia*	Gastropoda
33 D	Below	970	*Vermicularia*	Gastropoda
			Milleporidium	Stromatoporoidea
33 I	Across	935	*Vermicularia*	Gastropoda
33 K	Below	1100	*Astrocoenia*	Anthozoa (reef coral)
			Brachyseris	(reef coral)
			Cyathophora	(reef coral)
			Diploastrea	(reef coral)
			Microsolena	(reef coral)
			Montastrea	(reef coral)
			Actinostroma	Stromatoporoidea
			Stromatopora	Stromatoporoidea
			Pyrina	Echinoidea (sand dollar)

* Location on guyot with respect to break in slope
† Median depth of dredging in fathoms

Assemblages from Cape Johnson Guyot

37 A	Below	1050	*Astrocoenia*	Anthozoa (reef coral)
			Lophosmilia	(solitary coral)
			Montastrea	(reef coral)
			Nerinea	Gastropoda
37 C	Below	1085	*Brachyseris*	Anthozoa (reef coral)
			Caprina	Pelecypoda (rudistid)
			Cardita	Pelecypoda
			Cerithium	Gastropoda
			Trochus	Gastropoda
			Actinostroma	Stromatoporoidea

Coral.—The corals dredged from Hess and Cape Johnson guyots belong to the order Scleractinia, or stony hexacorals, and include six reef-building (hermatypic) and one solitary, or nonreef-building (ahermatypic) genera. Four of the seven genera are living on present-day reefs; three genera (*Microsolena, Cyathophora,* and *Brachyseris*) became extinct in the Cretaceous. Overlapping of the geologic ranges of the coral genera (Table 5) indicates an age between Aptian and Cenomanian.

The earliest occurrence of stony hexacorals was in the Middle Triassic. From that time to the present the hexacorals have increased in number of genera and species with several periods of "marking time" in spread and diversification. During the Mesozoic the regions of coral growth lay in the Tethyan Province: the regions covered by the world-encircling Tethys Sea (Suess, 1893, p. 183; Davies, 1934, p. 61): India, Middle East, Mediterranean, Caribbean, Texas, Mexico, and Central America.

In the Late Jurassic extensive reef building took place in the Tethys Sea and its extensions which included India, the Near East, the Mediterranean, England, and East Africa. Early Cretaceous time was a period of little or no reef building. During the Barremian and Aptian stages, however, another world-wide period of reef building began. Reef corals in these times extended from 47 N. Lat. to 10 S. Lat. (Vaughan and Wells, 1943, p. 69-75).

In Europe the Upper Barremian and Aptian are termed the "Urgonian". To quote from Wells (1933, p. 13):

> The Urgonian originally described by d'Orbigny as an horizon of the Upper Barremian is now understood to include the facies of the Upper Barremian and the Aptian represented by massive zoogenic limestones bearing a characteristic rudistid fauna and rarely ammonites, which occurs in several parts of the world and which is distinct from the littoral and bathyal facies. It may be considered as a special phase of the neritic facies characterized by warm, clear water fairly remote from land, typically reefs of corals or rudistids. (The Urgonian) appears in Mexico in the Lower Aptian and in Texas in the Upper Aptian.

The Trinity Group (Aptian) of Texas includes the coral genera *Astrocoenia, Montastrea, Cyathophora, Diploastrea,* and *Microsolena* (Wells, 1932, p. 225-256) all of which appear in the Mid-Pacific guyot area.

The Glen Rose formation of Texas (Lower Albian) contains an association of the reef coral genera *Astrocoenia, Cyathophora,* and *Diploastrea* (Wells, 1933, p. 10, 11). Wells (1933, p. 22) notes that in the Albian of Texas the reef corals did not produce any true reefs, but were colonies growing in scattered clumps amid the surrounding and underlying caprinids (pelecypods of the Family Caprinidae—also taken in the Guyot area).

The Urgonian reef-coral faunas appear in France, Italy, Spain, East Africa, Northern Venezuela, Texas, Mexico, and Japan. This world-wide spread was facilitated by the world-encircling Tethys Sea which then covered the present land bridges of the Near and Middle East, Central America, and Mexico. The currents of this great seaway probably ran from east to west and the free-swimming planula larvae of the coral could easily migrate in this direction.

Hess and Cape Johnson guyots are located between 17° and 18° N. Lat. The date of the fauna is Aptian to Cenomanian. In time, space, and faunal affinities the coral fauna appears to be that of the Tethyan Province and the closest geographical affinity appears to be with the Texas-Mexico area. The affinities of the individual genera are discussed in Appendix A.

The total aspect of the fauna is more that of the Urgonian (Upper Barremian and Aptian) than of other ages, even though the geologic range cannot be so restricted. This restriction would place the fauna in the Lower Cretaceous even though "Lower Cretaceous" has been variously defined. Some authors (Muller and Schenck, 1943, p. 271) do not use "Middle" Cretaceous and draw the lower boundary of the Upper Cretaceous below the Cenomanian. (See Table 5.) It is preferable, therefore, to use the stages of the Cretaceous (Aptian, Albian, etc.) in age discussions. The range Aptian to Cenomanian is nearer "Middle" Cretaceous, in general meaning, than Early or Late Cretaceous.

Fortunately for geologic dating of the probable submergence of the guyots, coral is restricted to rather definite ecologic conditions. The excellent paper on fossil and living coral by Vaughan and Wells (1943, p. 52-69) contains a concise summary of the conditions under which coral lives in modern seas. These are:

(1) Depth of water: The maximum depth at which corals are active in building reefs is about 25 fathoms, and most reef building takes place in depths of 15 fathoms, or less. Reef corals live between the surface and a maximum depth of 85 fathoms (Ladd *et al.*, 1950, p. 424). Solitary (ahermatypic) corals have been collected at a depth of 3209 fathoms.

(2) Temperature of water: Reef corals do not live in waters below a minimum temperature of 18.5° C (65° F). The optimum temperature lies between 77° and 84° F, with an average minimum of not less than 72° F. The maximum is around 97° F.

(3) Salinity: The salinity tolerated by reef corals lies between 27 and 40 parts per thousand; the average in regions of greatest development being about 36 parts per thousand. The salinity of the open Pacific Ocean is about 35 ppt. (Sverdrup *et al.*, 1946, p. 124).

(4) Exposure to air: Corals can withstand short intervals of exposure to air (as during lowermost tides) but not prolonged exposure.

(5) Sunlight: Sunlight is essential to coral reef growth. Apparently this is due to the need of the symbiotic algae in the coral tissue for sunlight for photosynthesis.

(6) Water circulation: Circulation of water to bring food, oxygen, and to remove sediment is necessary for strong coral-reef growth. This is the prime reason why the seaward side of coral banks is the favored spot for strong growth of coral (Vaughan, 1919, p. 214).

(7) Planula stage: The coral larva (the planula) exists in the larval stage from 2 to 30 days depending on the species. It is during this stage that coral migration takes place; the distance of migration depending on the currents as well as duration of larval stage.

(8) Distribution in present seas: Reef corals are found in the warm, shallow waters of the tropical and semitropical regions within the limits of 35° 10′ N. Lat. and 32° S. Lat. Reefs are restricted to the zone between 32° 30′ N. Lat. and 30° S. Lat.

In the study of paleoecology it is not always advisable to extend present-day ecologic conditions for any animal group too far into the geologic past. In the present samples, however, there are four genera of fossil reef-building corals which are living in the seas of today. Probable conclusions may therefore be drawn as to the ecologic conditions under which the fossil coral fauna of the guyots lived. These conditions may be summarized as follows:

The fossil coral lived in marine waters with a temperature not less than about 65° F and at a depth not exceeding 50–85 fathoms or 300–510 feet. The depth was probably less than 100 feet. The water was clear and circulation was sufficient to provide for food and removal of sediments.

The condition of fossilization is indicated by the fragmental nature of the pieces of coral, their integration as cemented constituents of larger rocks, and the coating of manganese dioxide on almost all specimens. These conditions, together with their present location below the break in slope, all argue for a detrital occurrence. The coral fauna is therefore considered to be an assemblage of animals brought together after death (a "thanatocoenose"). The location of the fossils on the guyot (below the break in slope) and their condition at the time of collection imply that the fragments had rolled a short distance down slope from living positions above and had formed the usual debris expected on the side slopes below a living coral reef or bank.

In view of the occurrence of reef coral with erosional debris it is a safe assumption that the dredge scraped off parts of the topmost layer of the hard material on the guyots; this evidence is supported by the fact that no younger fossils were dredged than the well-integrated-in-time megafossils. The death of the coral must then be explained. A reference to the ecologic conditions under which reef corals live reveals the limiting conditions which must be met to keep this coral alive.

A change in salinity of the Middle Cretaceous Pacific Ocean is excluded since later microfossils are exclusively marine and there is no known agency, or condition, which could have caused such a change. Sedimentation can cause the death of coral, but as the coral lived on the flat tops the major portion of the erosion had already taken place. The presence of a high, uneroded central portion of the guyots which later furnished sediment to kill the coral is unlikely since the coral banks would have protected such a central, high mass.

The temperature of the Cretaceous seas is known from widely separated studies to have been remarkably uniform and semitropical far north of the latitude (17°–18°) of the guyots. In addition the microfossils of the area in the latter part of the Cretaceous affirm that the area was still tropical to semitropical. It is unlikely, therefore, that a change in surface temperature of the sea caused the death of the coral.

A change in depth of water over the guyots to below that in which reef coral can live (about 50–85 fathoms) could have killed the coral. With the exception of some unknown ecologic factor this is the only cause of death left and by far the most likely one. The upward growth of coral is sufficient to keep pace with slow submergence and to keep the reef at the surface as the foundation relatively sinks. Submergence to kill coral must therefore be more rapid than the rate of growth of coral on its own debris. This is believed to be true in the guyot area of the Mid-Pacific Mountains.

It should be noted that the initially fast submergence postulated above need not submerge the coral farther than just below about 85 fathoms, the lower limit of survival of reef coral (Ladd et al., 1950, p. 424). Thus the first fast submergence might have been fairly small (50–85 fathoms) or even less if the coral lived in the lower part of its possible depth zone.

The death of the coral dates the initial submergence only (Aptian to Cenomanian) and has no bearing on the date of further submergence in the area to the present depth of more than a mile.

Submergence fast enough to kill coral is indicated by many examples listed by Davis (1928, p. 473–510). These submerged reefs range in form from banks to apparently drowned atolls. The Dutch Siboga Expedition dredged manganese dioxide-coated reef coral in the Ceram Sea at a depth between 711–893 fathoms (Davis, 1928, p. 494). Molengraaff thought the coral was from a reef submerged by downwarping in the area.

Rudistids.—"Rudistid" is a common name for several families of pelecypods which had an aberrant growth form and lived in the marine waters of Jurassic and Cretaceous time. One family of rudistids (Caprinidae) is characterized by wide, simple, parallel canals in the middle layer of the upper, and in some cases, the lower valve. All the rudistids dredged from the Cape Johnson Guyot appear to belong to the family Caprinidae and to the genus *Caprina*, which would date the fauna as Aptian to Turonian (Haug, 1909, p. 1169; Palmer, 1928, p. 58; *See* Table 5).

Paquier (1903, p. 71) believed that the nonbifurcating canal walls (true of the guyot caprinids) are more primitive than the bifurcating type. Palmer (1928, p. 21) noted that no bifurcations of the canal walls have been noted below the Cenomanian and subsequent to that stage the simple plates were rare. The genus *Caprina* occurs sparsely below the Cenomanian. Its rapid expansion is confined to the Cenomanian, although the genus is reported in the Turonian (Haug, 1909, p. 1169).

One of the species of *Caprina* taken on Cape Johnson Guyot is closely allied to the genus *Praecaprina* Paquier which was the primitive caprinid dated as Aptian and believed by Paquier (1903, p. 81, 82) to be the ancestor of *Caprina*. This relationship suggests an age for the caprinid material from Cape Johnson Guyot nearer Aptian than Turonian.

Two other caprinid fragments tend to support a geologic age early in the range Aptian to Turonian. One shell fragment has canal structure which resembles that of *Caprina choffati* described by Douville (1898, p. 143–147) from the Albian of Portugal. A second shell apparently is close to *Diceras* and probably belongs to the same family, which lived from Late Jurassic to Early Cretaceous time.

In view of the above evidence the Cape Johnson rudistid fauna may be restricted to the geologic range Aptian to Cenomanian which agrees with the age of the coral fauna discussed previously.

Palmer (1928, p. 18) showed on a distribution map that rudistids occupied the exact geographical limits of the Cretaceous Tethys Sea. In this area these faunas were closely associated with reef corals. The genus *Caprina* is known from Southern England, the Mediterranean area, Texas, and Mexico, thus the affinities of the rudistid fauna are with those of the Tethys Sea of Cretaceous time (Aptian-Cenomanian) in these areas.

The genus *Caprina* and all other rudistids are extinct so their ecology cannot be surmised from that of modern forms. However, convincing indirect evidence, is available for such postulations. In all regions of caprinid occurrences these forms lived in close association with reef corals, so that this fossil community is commonly referred to as a "reef coral-rudistid" or "reef coral-caprinid" fauna. Therefore it is a safe inference that caprinids must have lived in the same environments as reef coral: warm, shallow, clear marine waters of the tropical and semitropical regions.

All the caprinid material came from MP 37 C just below the break in slope on Cape Johnson Guyot. The fragments came from detrital rock which included fossils, indurated globigerina ooze, fragments of limestone, and crusts and nodules of manganese dioxide all cemented together. Some of the caprinids have been phosphatized. This occurrence appears to argue for an assemblage as an accumulation of dead organisms (thanatocoenose). The caprinids, along with the corals, are thought to have rolled down slope from growth positions above. Prior to their dislodgement the caprinids, corals, and stromatoporoids probably were a normal association of living organisms (biocoenose).

Stromatoporoids.—Stromatoporoids are extinct members of the phylum Coelenterata. Their exact taxonomic position is uncertain, but they are thought to be related to Hydrozoa (Shimer and Shrock, 1944, p. 58; Swinnerton, 1947, p. 32; Woods, 1950, p. 76). Some authors give them a group or class systematic position in the phylum Coelenterata, although others include them as an order in the class Hydrozoa. In the early Paleozoic these organisms built extensive calcareous reefs in which some of the individuals attained a size up to 5 feet in diameter. Many stromatoporoids consist of concentric, undulating, calcareous laminae through which radial pillars may extend.

For many years paleontologists thought that stromatoporoids became extinct at the end of the Paleozoic. There are, however, numerous occurrences of stromatoporoids in the Mesozoic where they finally became extinct in the Cretaceous. Some of the papers dealing with Mesozoic stromatoporoids are by Yabe and Sugiyama (1935), Dehorne (1920), Parona (1932), Wells (1934c), Steiner (1932), and many others.

Three genera of stromatoporoids were dredged on Cape Johnson and Hess guyots. *Milleporidium* and *Stromatopora* were taken on Hess Guyot, and the same species of *Actinostroma* was taken on both guyots. The geologic range of *Milleporidium* is Triassic to Cretaceous (Danian); *Stromatopora* ranges from Silurian to Cretaceous (Cenomanian); and *Actinostroma* from the Cambrian to Cretaceous (Senonian). These genera, then, are important evidence for a Pre-Tertiary age of the guyot faunas.

Stromatoporoids are thought to have been warm-, shallow-water animals because of their associations with reef corals and rudistids. An example of this association occurs in the Torinosu limestone (Upper Jurassic) of Japan which contains stromatoporoids, reef coral, echinoderms, mollusks, and calcareous algae (Yabe and Sugiyama, 1935, p. 139).

All of the stromatoporoids dredged from the guyots were detrital material taken below the break in slope. On Hess Guyot they were intermixed with reef-coral fragments; on Cape Johnson Guyot they occurred with fragments of rudistids and reef coral. The stromatoporoids are considered to have died and rolled down slope where they contributed to the reef detrital material.

Gastropods.—Three genera of the class Gastropoda (*Nerinea, Cerithium,* and *Trochus*) were dredged from Cape Johnson Guyot. All three of these genera have been reported as constituents of the Cretaceous reef coral-rudistid faunas. *Nerinea* which became extinct in the Upper Cretaceous (Senonian) is the only genus important in dating the assemblage as Cretaceous. Felix (1891, p. 142–172) recorded *Nerinea*, reef coral, and rudistids from the Lower Cretaceous of Mexico. Blankenhorn (1927, p. 114) reported *Cerithium, Nerinea,* and rudistids from the Cretaceous (Cenomanian) of Palestine, and *Trochus, Nerinea,* and *Cerithium* from the Cenomanian of Syria.

Echinoid.—The dredge haul of coral (MP 33 K) from Hess Guyot also yielded an excellently preserved specimen of the genus *Pyrina* of the class Echinoidea, order Irregularia (Exocyclica). *Pyrina* was a characteristic Cretaceous genus but it also occurred sparsely in the Jurassic and has been reported from the Eocene of Italy (Dames, 1877, p. 18). The genus occurred abundantly in Europe, especially in France where it appeared in the Early Cretaceous, reached a maximum in the Senonian, and died out by the end of the Cretaceous (d'Orbigny, 1853–1860). At least one species occurred in the Cretaceous of North America (Cooke, 1946, p. 221). Kuhn (1933, p. 151) recorded *Pyrina* from the Senonian of Persia. Neaverson (1928, p. 413) recorded the following information concerning *Pyrina*:

—*Pyrina*, of which *Echinoneus* is the living homeomorphous derivative. The latter lives in sheltered places near low tide mark and is essentially a shallow water form. *Pyrina* is considered to have had the same general mode of life and its highest occurrence in this country (Britain) is in the Chloritic Marl (Cenomanian), the last well defined littoral deposit of the English Cretaceous. In France, however, the genus persists into the Senonian, but only where the deposit is full of detrital matter. Thus *Pyrina* is confined to littoral tracts.

Kuhn (1933, p. 151) reported *Pyrina* as a member of an Upper Cretaceous fauna which included reef coral, rudistids, and mollusks. Kirk and MacIntyre (1951, p. 1505; and personal communication) report *Pyrina* with reef coral and rudistids from the Cenomanian of Baja California, Mexico.

In summary, *Pyrina* lived in shallow water and was a member of the reef coral-rudistid-stromatoporoid association of Cretaceous time.

Miscellaneous.—At MP 34 on the northern side of the main ridge of the Mid-Pacific Mountains a solitary, nonreef-building coral was dredged from the upper slopes of a sharp, apparently uneroded peak. This coral is very poorly preserved but seems to fall into the subfamily Oculininae, a group of ahermatypic corals which ranges from Tertiary to Recent. This solitary coral has no apparent time relation to the guyot faunas farther south and is not indicative as to depth.

A fossil shark's tooth of the genus *Oxyrhina* (?) was dredged from Guyot 19171 (MP 26 A-3). The genus ranges from Early Cretaceous to Recent, but fossil shark's teeth cannot be accurately used in dating (Durham, Personal communication to R. S. Dietz).

Coralline algae were the only plant life dredged from the guyots. These algae came up with the coral from Hess Guyot. Because algae are restricted to the light zone their occurrence is important evidence of "shallow" water environment. Algae live to depths of 60 fathoms or less (Gardiner, 1931, p. 64). No attempt was made to identify the coralline algae; by association with the coral of the same dredge haul its age should be Cretaceous.

The earbone of a cetacean (Vanderhoof, Personal communication) was dredged from Cape Johnson Guyot. These fossils are not diagnostic as to age or depth. Many were collected by the Challenger Expedition (Murray and Renard, 1891, p. 270).

Conclusions.—The megafossils of Hess and Cape Johnson guyots form a well-integrated reef-coral, rudistid, stromatoporoid, echinoid, and molluskan fauna which lived in the Cretaceous Period from Aptian to Cenomanian (Turonian?) time. The affinities of the fauna are with those of the Tethys Sea to the east.

The ecologic conditions implied are: marine waters of average salinity; shallow waters not exceeding 50–85 fathoms in depth; clear water relatively free of sediment; tropical to semitropical waters.

The sedimentary rocks of the guyots are Cretaceous, the oldest yet to be discovered in the Pacific Basin. The shallow-water fauna shows that the guyots were once at or near the surface of the Pacific, and the Mid-Pacific Mountains once formed a chain of islands. Submergence sometime during Aptian to Cenomanian time killed the coral.

The Cretaceous Pacific Ocean was at least as deep as the present height of the Hess and Cape Johnson guyots (about 1700 fathoms) and it can be concluded that the Pacific Ocean was present and occupied a deep basin at least as far back as the Cretaceous.

The Tethyan Faunal Province is now extended several thousand miles into the Northern Pacific and the long distance trans-oceanic migration of many sessile benthonic and shallow-water animals is indicated.

The data do not support the speculations of Hess (1946, p. 789) that guyots are probably free of coral and were Precambrian islands. However, the evidence of this paper concerns only five guyots in one general locality and does not purport to solve the problem of *all* guyots.

MICROFOSSILS

General discussion.—The planktonic Foraminifera dredged and in cores from the guyots are discussed systematically and illustrated in a separate paper (Hamilton, 1953), but a summary is presented in this paper for completeness. Appendix A gives detailed faunal lists.

Cores at MP 27 and 27-2P contained a mixed assemblage of Upper Cretaceous to Recent planktonic Foraminifera. The Upper Cretaceous fauna contained 12 species of the genus *Globotruncana* which is restricted to the Upper Cretaceous, 7 species of the genus *Gümbelina*, and 2 species of the genus *Ventilabrella*. The Cenozoic fauna ranged through the Tertiary and included the modern tropical Pacific planktonic forms.

At MP 33 C on Hess Guyot an Upper Paleocene (Upper Midway) foraminiferal fauna occurred in the indurated globigerina ooze (Pl. 2, fig. 3) that was dredged from about 915 fathoms near the center of the guyot. This fauna was dominated by variations around the species *Globorotalia velascoensis* Cushman and contained several species of *Gümbelina*.

Two cores (MP 25 E-1 and E-2) taken just above the break in slope on Horizon Guyot contained the same Lower Eocene fauna in white, unindurated globigerina ooze. The top of both cores was a mixture of Eocene and Recent species. An uncontaminated Lower Eocene planktonic fauna occurred an inch or two beneath the surface in both cores. This fauna contained *Globorotalia crassata* (Cushman), *G. aragonensis* Nuttall, *Hantkenina mexicana* Cushman var. *aragonensis* Nuttall, and *Hastigerinella eocanica aragonensis* Nuttall; the latter three species are distinctive constituents of the fauna of the Aragon formation of the Tampico Embayment Region of Mexico.

A dredge haul (MP 26 A-3) across the break in slope on Guyot 19171 brought up a rounded boulder of indurated globigerina ooze which was completely covered by a manganese-dioxide coating which was 6 mm thick on top but merely a black discoloration on the bottom. This boulder contained a Lower Eocene planktonic assemblage the tests of which were affected by solution and reprecipitation.

Early Tertiary species were noted in all of the cores which were composed mostly of present-day species of tropical planktonic Foraminifera.

Age, affinities, and ecology.—The Cretaceous planktonic foraminiferal fauna of MP 27 is a well-integrated Campanian-Maestrichtian assemblage and is very closely related to faunas of the same age from the Gulf Coast, Caribbean, and adjacent areas (Upper Taylor-Lower Navarro; Cushman, 1946, p. 9–13). It is a typical assemblage from the tropical to semitropical Tethyan Province.

In core MP 27-2 layers of sand and gravel are separated by globigerina ooze and red clay in which the constituent grains are both graded and ungraded. The bottom layer of gravel in MP 27-2 is definitely graded. As discussed more fully in the separate paper on the Foraminifera (Hamilton, 1953) the gravel probably arrived in the basin near the bottom slopes of a near-by guyot after transportation by turbidity currents. This is supported by the graded gravel layer and the mixed Upper Cretaceous to Recent foraminiferal fauna.

Before slumping or sliding on the slopes of a nearby eroded seamount there would have been continuous deposition of normal faunas. If the tops and upper side slopes of the guyots are essentially areas of nonaccumulation then the results might well have been a condensed bed with a "condensed fauna" (a bed containing all elements of the faunas but not as thick as it would ordinarily be). The Cretaceous fauna found in the gravel layers was probably originally the planktonic fauna which was deposited with the gravel. When the submarine landslide, or slump, began to move and turned into a turbidity current, the whole mass of sediment together with its foraminiferal fauna became mixed. This mixture of Cretaceous to Recent species settled out in the basin to form the mixed fauna obtained in cores at MP 27.

The anomalous fact that restricted Upper Cretaceous species are found at the surface of the present sea bottom may be explained in at least two ways: (1) the postulated sequence of events occurred quite recently, and normal planktonic deposition has not buried the newly deposited material, or (2) the events occurred in the geologic past, and the basin of deposition is now an area of nonaccumulation owing to deep currents The configuration of the Mid-Pacific Mountains in this area might restrict and funnel currents through the basin, although such deep currents (at 2000 fathoms) are unknown.

The planktonic assemblage from the indurated ooze of Hess Guyot appears to be Upper Paleocene (Upper Midway) on the basis of overlapping geologic ranges of its species. It is most closely related to the Velasco shale of the Tampico Embayment Region of Mexico.

The fossil planktonic assemblage from Horizon Guyot, which is probably late Early Eocene, is very close to the planktonic assemblage from the Aragon formation of the Tampico Embayment Region of Mexico.

The Foraminifera, quite independently of the megafossils, date the erosion of at least one guyot as pre-Late Cretaceous (fauna in the gravel at MP 27), and date the flat top of one guyot as pre-Late Paleocene, and those of two other guyots as pre-Early Eocene.

That the Tethyan Province of the Cretaceous and Early Tertiary was tropical to semitropical is known by the occurrence of reef coral. The genera *Globotruncana* and *Hantkenina* apparently were confined to the Tethyan Province and adjacent areas and can thus be considered to have been Foraminifera of the then tropical to semitropical waters. The occurrence of these genera on the Mid-Pacific guyots affirms that the area was still covered by warm waters in the Late Cretaceous and Early Tertiary as it was when the reef coral-rudistid fauna grew on the tops of the ancient islands earlier in the Cretaceous.

The discovery of fossil planktonic index species of Foraminifera from the flat-topped seamounts of the Mid-Pacific places on record the oldest foraminiferal assemblages yet discovered from the Pacific Basin and extends the geographical ranges of many planktonic Foraminifera previously recorded from the Middle East and the Gulf and Caribbean areas several thousand miles into the Pacific Basin.

The writer has already discussed the species whose geographical ranges have been thus extended and presents geologic-range charts of the important species identified (Hamilton, 1953).

It is now possible to construct a range chart of the index fossils of planktonic Foraminifera for the whole tropical and semitropical Pacific Basin. In view of the fact that the location of these fossil oozes was in the Middle Pacific, and the fact that they were planktonic assemblages drifted by the currents, it is a safe conclusion that these fossil Foraminifera were present during their lifetime all over the tropical and semitropical Pacific Basin. This is not surprising, nor unusual, but this is the first report of the occurrence of most of these forms from the deep Pacific Ocean. Their presence could have been logically postulated and predicted, but it is satisfying to know for a fact that they were actually deposited and that they are available to date Pacific sediments and to help solve problems of sedimentation in future work.

LITHOLOGY AND SEDIMENTS
GENERAL

The rocks and sediments were examined in the hand specimen, in 40 thin sections under the petrographic microscope, by mechanical and spectrographic analysis, and by insoluble residues. The detailed descriptions of most of these examinations are in Appendix B.

IGNEOUS ROCKS

Igneous rocks collected on the guyots are summarized as follows:

Guyot	Location on the guyot	Type of rock
Horizon	Just below the break in slope (MP 25F-1 and F-2)	Highly altered volcanic rock; unaltered olivine basalt; subrounded olivine basalt pebbles
19171	Across break in slope (MP 26 A-3)	Large rounded boulders of olivine basalt with rounded corners; subangular pebbles of olivine basalt
20171	Just below break in slope (MP 26)	Subrounded basalt pebbles
Core	MP 27-2 below slopes of Guyot 20171	Subangular to rounded pebbles of basalt and olivine basalt as layers of gravel in red clay
Hess	Near center of top	One well-rounded pebble of olivine basalt
Cape Johnson	Just below break in slope	Sand grains of basalt from large sandstone boulder

According to modern hypothesis the earth's crust under the Pacific Basin is composed of basaltic material (Daly, 1942, p. 59; Kuenen, 1950, p. 115–123). This hypothesis is supported by geophysical observations and several lines of theoretical reasoning. Kuenen (1950, p. 117) divides the supporting evidence into three parts; seismic, gravimetric, and petrographic. The findings of the present study add only to the petrographic evidence.

The basaltic rocks of the Pacific Basin may be separated from the landward, more acidic rocks by an "andesite line" (Hobbs, 1944, p. 262; Hess, 1948, p. 422). The andesite line runs from Alaska to Japan, the Marianas, Palau Islands, Bismarck Archipelago, south to the east of New Zealand via the Fiji and Tonga groups. On the east side of the Pacific Basin the andesite line runs along the continental margin of the Americas. Carsola and Dietz (1952, p. 492) note that the line probably runs between Fieberling Guyot and the islands off the California coast. All the Mid-Pacific Mountains, therefore, lie well within the basaltic area included within the andesite line.

Within the andesite line in the north Pacific Basin there are two kinds of islands: the low coral atolls which have yielded only calcareous rocks from organic sources, and volcanic islands. The petrology of the Pacific Islands has been extensively studied (Daly, 1916; Washington, 1923; Washington and Keyes, 1926, 1927, 1928; Barth, 1931; Macdonald, 1949b). In all areas this volcanic rock is dominantly basalt; olivine basalts are by far the dominant type with minor amounts of picrite-basalt, basalt, and trachyte making up the major portion of the remaining types. The olivine basalt suite has been named the Hawaiian type basalt.

The igneous rocks dredged from the guyots of the Mid-Pacific Mountains are of the olivine basalt Hawaiian suite and furnish another locality within the andesite line which yields additional evidence of the validity of the hypotheses concerning the rock types to be expected within this province.

These basalts are mostly erosional debris: sand, pebbles, cobbles, and boulders which range from angular to well rounded. Most of this material was dredged just below the break in slope where erosional debris would be expected to accumulate.

The amount and location of the basaltic debris precludes such exotic explanations as transportation by ice rafting, in the roots of plants and trees, or in the stomachs of seals or sea lions.

Dietz and Menard (1951, p. 2011) summarize the evidence and opinions on the depth at which waves and currents will abrade hard rock. Umbgrove (1947, p. 104) estimated a depth of about 40 fathoms for the limit of wave abrasion; Barrel (1917, p. 779-780) estimated 50 fathoms. Dietz and Menard place wave abrasion in the zone of vigorous wave action and consider that only slow abrasion takes place below about 5 fathoms. They point out that the formation of an abrasional platform with an abrupt marginal break of slope requires strong currents down to platform depth, and weak currents directly below it This condition prevails only near the ocean surface where breakers form.

The definite signs of wear and rounding of the basaltic debris prove that the guyots were eroded, and only vigorous wave action near the surface could have been the cause.

The presence of basaltic erosional debris near the tops of the guyots, and in one place near the center of the top, indicates that the guyots are not sunken atolls. A modern atoll completely sheaths its supposed foundation with thick layers of organic limestone and debris of coral reefs; the side slopes near the top are completely covered by talus accumulations from the reefs above. No basalt debris could be dredged from the uppermost slopes of a true atoll.

In one dredge haul reef coral and sandstone from a basaltic source were taken together from just below the break in slope. A well-rounded pebble of basalt was dredged from near the center of Hess Guyot, further evidence against the guyots' being true atolls. The coral and rudistids apparently grew in banks on and among the erosional debris The organic material was insufficient to conceal the eroded material.

LIMESTONE AND CALCAREOUS OOZES

General.—Limestone and calcareous oozes occurred on the guyots as follows:

Guyot	Location with respect to break in slope	Type
Horizon	Above-on top	Globigerina ooze cores: 28 and 6 inches
	Just below	Globigerina ooze cores: 28 and 6 inches; indurated calcareous ooze as cement of broken volcanic rocks and manganese
Guyot 19171	Above	Indurated globigerina-ooze boulder
Guyot 20171	Just below	Globigerina ooze from dredge
Hess	Just above	Limestone of rounded pebbles and pelecypod shells cemented by calcite; globigerina ooze
	Above-on top	Coquina of gastropod shells and calcareous ooze; indurated globigerina ooze; globigerina ooze cores up to 31 inches; dredged globigerina ooze
	Just below	Organic limestone of reef coral and algae
Cape Johnson	Below	Globigerina ooze in dredges and in cores up to 108 inches; indurated calcareous ooze; fossils
	At break	Globigerina-ooze cores up to 26 inches

Globigerina ooze.—Shepard (1948, p. 303), Ladd and Tracey (1949, p. 3), and others, have stated that fine sediments do not accumulate on topographic highs on the sea floor due to removal by waves and currents. The findings of Mid-Pac reveal that this is essentially true on the Mid-Pacific guyots. Although globigerina ooze was sampled on all five of the guyots, the quantity was not commensurate with the time these bottom features have been submerged.

The tops of the guyots are essentially areas of nonaccumulation because currents remove the fine pelagic sediments. Evidence for this statement comes from two sources. The present flat tops could not have been preserved had pelagic sediments been allowed to pile up unhindered since Cretaceous time. In such a case the guyots would appear as low, conical peaks, the topmost part being a huge mound of globigerina ooze. Furthermore, Paleocene and Eocene Foraminifera are at the surface on Hess and Cape Johnson guyots (mixed with Recent fauna) which implies that ancient accumulations of globigerina ooze are exposed and their erosion-furnished fossil Foraminifera which are being redeposited with the slow rain of present-day planktonic Foraminifera. The pure, unconsolidated Eocene globigerina ooze an inch or two beneath the surface of Horizon Guyot is particularly significant. These deposits of Early Tertiary oozes may have accumulated initially in hollows, small erosion channels, and among the rocks. When the bank was essentially level, currents kept the top as an area of almost no accumulation. Thus the date of fossil sediments at or near the surface may indicate the time when the bank became essentially flat, practically ceased to accumulate sediments, and planktonic forms deposited later were drifted off the guyot by weak currents. Occasional slumps probably have occurred on the side slopes and on the tops of the guyots, especially near the edges, and might have been an important factor on Horizon Guyot where a pure Lower Eocene globigerina ooze occurs just beneath the surface which is composed of Eocene and Recent species. Extensive deposits of globigerina ooze on the guyots indicate that there is no present-day erosion of the basaltic platform (a criterion mentioned by Kuenen, 1950, p. 231).

Two pictures (Menard, 1952, p. 5, 7) taken on Sylvania Seamount, a guyot just northwest of Bikini Atoll, probably illustrate the typical occurrences of globigerina ooze on a guyot. One shows an unbroken expanse of ooze. The other shows globigerina ooze on and among manganese-coated rocks. The presence of oscillatory ripple marks at a depth of 745 fathoms indicates that deep submarine water movements exist. Menard (1952, p. 8) suggests that internal waves may be a possible agent in forming these deep ripples.

Indurated globigerina ooze was dredged on four of the guyots. Murray and Lee (1909, p. 22) record three instances (one each by the ALBATROSS, CHALLENGER, and BRITANNIA) in which hardened globigerina ooze was dredged from the sea floor. On the Mid-Pacific guyots the material is dry, hard, calcium carbonate ooze cemented by calcite and usually coated, at least in part, by manganese dioxide. On Hess Guyot the material was Paleocene and on Guyot 19171 it was Eocene in age. Evidence at hand is insufficient to explain the occurrence of this hardened ooze but a possible explanation might be compaction under overlying sediments, followed by reduction in water content and precipitation of calcium carbonate as a cementing material. Later removal of the unconsolidated overlying material would leave the hardened material at the surface where it could become coated with manganese dioxide. Also the dredge could have removed pieces of hardened ooze from under a cover of unconsolidated material.

Detrital limestone.—The detrital limestone from Hess Guyot (MP 33 A), composed of rounded, calcareous pellets and pelecypod shells cemented by calcite (Pl. 4, fig. 2), resembles the beach limestone formed on the shores of modern atolls. A bed of this material, however, could have been formed by solution and reprecipitation of calcite under water in compacted beds and an ancient beach is not necessarily implied.

Silicified Limestone.—Silicified limestone was detected in manganese nodule centers on two of the Mid-Pacific guyots (MP 25 and MP 26). Silicified limestone from the deep-sea bottom is rare, but has been previously reported as centers of manganese nodules (Murray and Lee, 1909, p. 28). A manganese nodule from Horizon Guyot (MP 25) was chemically analyzed at Scripps Institution (Patterson *et al.*, 1953, p. 1388) and was determined to have 12.1 per cent of nearly pure silica in the form of a spongy network.

Silica is present in sea water in true ionic solution and not as a colloid (Roy, 1945, p. 401). The possible sources of silica in unusual amounts in the Mid-Pacific can be confined to deposits of siliceous organisms such as diatoms, the submarine solution of volcanic rocks, or siliceous magmatic emanations such as volcanic fluids and gases, or volcanic springs. There are no known deposits of diatoms or radiolaria in the area from which silica could be redistributed. Diatoms and radiolaria live in the tropic and semitropical waters but do not collect in significant deposits unless, owing to depth or other causes, the calcareous oozes are absent.

The association of siliceous sediments and volcanic material is well known (Lawson, 1895, p. 420–426; Rubey, 1929, p. 166–169; Taliaferro, 1933, p. 54; Bramlette, 1946, p. 37–57). Bramlette thinks that the source of the silica for the extensive diatomaceous deposits of the Monterey formation was the Miocene pyroclastic rocks. He stated that one of the earliest and quantitatively most important effects of the alteration of vitric pyroclastics is a loss of silica; great quantities of silica would be dissolved in sea water from volcanic ash.

The dying phases of volcanic activity are fumarole and hot springs activity. These fluid and gaseous emanations are high in silica and manganese. In view of the basaltic nature of the Mid-Pacific Mountains the volcanic sources (pyroclastics or emanations) are distinct possibilities for the supply of silica which replaced the organic limestone on some of the guyots. Metasomatism is the process indicated by the replaced calcareous material. Limestone is an especially favorable host rock for such replacement (Emmons, 1940, p. 160).

Phosphatized limestone.—Phosphatized limestone occurred as subrounded to subangular centers of manganese nodules on three of the guyots. On Cape Johnson (MP 37 C) about 200 pounds of fossiliferous, manganese-coated and partially phosphatized limestone was dredged. In thin sections the organic source of the limestone is indicated by the phosphatized tests of Foraminifera.

When the phosphatized rock is dissolved in nitric acid, and ammonium molybdate is added, a yellow precipitate is formed which indicates the presence of phosphate. The extent and definitive proof of the phosphatization was furnished by Robert W. Rex of Scripps Institution of Oceanography in special X-ray powder studies of the rocks from Pacific seamounts. Additional chemical tests through the courtesy of E. D. Goldberg of Scripps confirmed these studies.

Two representative samples of partially phosphatized globigerina ooze were analyzed spectographically by C. E. Harvey of the American Spectographic Laboratories with the following results:

Elements as oxides	Station	
	MP 26 A-3 (center of nodule) per cent	MP 37 C Cape Johnson Guyot per cent
Calcium CO_2 and anions	65.5	61.2
P_2O_5	28.0	35.0
SiO_2	1.9	1.2
Other	4.6	2.6

The phosphatized limestone from Cape Johnson guyot appears as dark-reddish, irregular phosphatic slabs, flakes, and veins in the unaltered, indurated calcium carbonate ooze. Many of the calcareous megafossils were partly or wholly phosphatized.

Phosphorite deposits on the sea floor are well known. Murray and Renard (1891, p. 391–400) note the presence of phosphatic material in both deep and shallow waters and in one instance the phosphate had replaced globigerina ooze (1891, p. 394). Other sea-floor occurrences are noted by Murray and Lee (1909, p. 38), Dietz et al. (1942), and Emery and Dietz (1950). Dietz, et al. (1942) discuss the distribution, origin, chemistry, etc., of phosphorite on land and in the sea, and it is unnecessary to review the subject here. They conclude that calculations indicate that ocean water is essentially saturated with tri-calcium phosphate and that slight changes in the physical-chemical environment might be expected to cause the direct precipitation of tri-calcium phosphate, but that the material actually precipitated is a compound having the apatite structure and that collophane is the principal mineral.

Emery and Dietz (1950) stress the abundance of phosphorite in nondepositional areas such as offshore banks, steep slopes, and the outer edge of the shelf near a break in slope. The guyots belong to this same general type of essentially nonaccumulation environment.

MANGANESE DIOXIDE

Manganese dioxide was present on all of the guyots where it formed coatings and crusts on all types of rocks and sediments and formed concentric layers around centers of silicified and phosphatized limestone, basalt, aggregates of globigerina ooze, and a shark's tooth. Figure 4 of Plate 4 shows typical coating. A manganese nodule is illustrated in Figure 3c of Plate 4.

The occurrence of manganese dioxide is extremely widespread in the present-day sea bottom and is reported by all deep-sea sediment-sampling expeditions. The first comprehensive reports were by Murray (1877, p. 247) and Murray and Renard (1891, p. 341–378) who described the sediments collected by HMS CHALLENGER. The manganese of the ocean floor is mineralogically close to "wad," the impure manganese ore, but the lattice structure of the nodules is dissimilar to any known iron or manganese mineral (Goldberg, 1954, p. 257). Goldberg (1954, p. 252) has made six spectrophotometric analyses of manganese nodules from Cape Johnson Guyot and MP 26A-3 with the following results (averaged at each station):

Station	Fe	Mn	Fe/Mn	Ni	Co	Cu	P	Al	Ti	Zr
					Weight per cent					
MP 26A-3	16.8	16.9	1.00	0.31	0.46	0.19	1.23	0.27	1.25	0.0057
Cape Johnson Guyot	13.4	21.0	0.64	0.58	0.54	0.47	0.54	0.39	0.81	0.0056

Until recently very little research had been done on sea-floor manganese. Several writers have called attention to the association of manganese oxides and volcanic rocks and pyroclastic material (Murray, 1877, p. 247; Clarke, 1920, p. 135; Park, 1946, p. 305; Goldberg, 1954, p. 259). Clarke (1920, p. 135) believed that manganese is easily derived from the alteration of volcanic-rock fragments, that it goes into solution as carbonate, is oxidized by the dissolved oxygen of sea water, and is precipitated near its point of derivation around any nuclei at hand.

The most thorough recent study of the process by which manganese and iron are concentrated in marine sediments as coatings around nuclei, or on rock, was made by Goldberg (1954; and personal communication). Goldberg proposes an electrochemical process for the concentration of nearly equal amounts of iron and manganese in nodules as a result of the scavenging action of iron oxides. Mass water movements, such as tides, are proposed by Goldberg (p. 251) as the energy source for the process of plating out the manganese and iron. Goldberg's hypothesis furthermore explains the concentration of the other elements noted in the analyses.

Park (1946) discussed the close association on the Olympic Peninsula of red, siliceous limestone,, manganese, and basalt flows. He noted that ordinary basalt contains 0.17–0.31 per cent MnO and that assuming only 0.01 per cent of Mn removed, the alteration of a cubic mile of basalt of specific gravity 2.90 would yield about 1,333,980 tons of MnO, or a basalt flow 1 mile square and 100 feet deep would yield about 25,264 tons of MnO. Park explains the Olympic Peninsula associations as follows:

(1) The lava accumulates rapidly in flows under water.

(2) Some flows break up into pillow lavas with thermal agitation aiding limestone deposition in overlying waters; volcanic debris is deposited with the limestone.

(3) Upon cessation of lava extrusion, the lava pile is covered by a thick accumulation of limestone and finely divided volcanic debris.

(4) Within the pile and beneath the cap (3), heat is retained; steam and aqueous solutions derived in large part from sea water trapped in interspaces, or distilled from underlying muds, would circulate freely; primary gases from crystallizing lavas would mingle with the heated aqueous solutions.

(5) The results of these steps would be: (a) albitization of labradorite; (b) solution of the femags of basalt including manganese; (c) enrichment in silica by dissolving from muds and particularly from the decomposition of femags and other silicate minerals.

(6) Solutions rich in silica, manganese, iron, and calcium would migrate upward toward the top of the volcanic pile; when the solutions came in cooler zones near the top, silica would be deposited as jasper and locally replace tuffs, limestone, or

lava; manganese combined with silica would be precipitated or replace limestone; large quantities of calcium ion set free from the spilite reaction might be a factor in the precipitation of more limestone.

This sequence of events could have taken place in the area of the guyots and it should be noted that this hypothesis explains the production of siliceous limestone, the inorganic precipitation of limestone, and manganese accretion. Evidence that sharp, uneroded peaks are nearer the surface of the sea than are the flat tops of the guyots may indicate that volcanic activity did not cease in the area of the Mid-Pacific Mountains after the flat tops of some seamounts had been truncated and part or all of the range had subsided. Continuing volcanic activity in the area might have furnished emanations and solutions which, with the submarine solution of pyroclastics, caused local concentrations of silica and manganese in the bottom waters.

The concentric layering of the nodules (Pl. 4, fig. 3c) is evidence for a concretionary origin and for the suspension of these nodules in the sediments where accretion could take place on all sides. Nodules formed around nuclei on a hard surface would tend to be flat on one side; otherwise a constant rolling of the growing nodule would be necessary to produce the concentric form, an unlikely event except in areas where strong currents moved over the bottom.

GEOMORPHOLOGY
INTRODUCTION

From their relationship with each other and with the Mid-Pacific Mountains, from the morphology, rocks, sediments, and paleontology of the guyots, deductions can now be made as to what land form is represented by the guyots and as to what kind of mountains are represented by the Mid-Pacific Mountains.

Johnson (1933) outlined in his "Analytical Method" the ideal prosecution of field work and the correct use of the "Theory of Multiple Working Hypotheses" (Chamberlin, 1897). It should be apparent to the "dry-land" field geologist that work in submarine geology presents unusual difficulties in scientific investigations and in the use of Johnson's method. A submarine geologic exploration into the deep-sea basins is usually a "one shot" investigation requiring large scientific parties and the use of exceedingly expensive equipment (including ships and their crews).

Any scientific field work includes gathering data, evolving hypotheses, deducing consequences of these hypotheses, and then confronting these deductions with actual findings in the field, which usually requires at least a second period of field work. This second period of field work is usually impracticable in submarine geologic investigations. That Johnson's method can be used, but with stringent limitations, is illustrated by the exploration of the Mid-Pacific guyots.

The Mid-Pacific Expedition intended to explore, dredge, and core previously located guyots for sediments, rocks, and if lucky, fossils; no time was allowed for general, areal exploration. The possible rock and fossil types which might be found (including the correct one of reef coral) were discussed prior to the actual exploration. The various landforms (truncated volcano, drowned atoll, sediment-filled caldera or crater, etc.) and hypotheses to be induced from these sediments and landforms were considered; and deductions made as to the consequences of these hypotheses. These deductions guided the sampling procedure eventually followed. Sampling across the break in slope between the flat top and side slopes was stressed, because of such reasoning by the Sea-Floor Studies Section.

GEOMORPHOLOGY OF THE GUYOTS

Previously discussed evidence leaves a number of working hypotheses for the landform represented by the coral-bearing, flat-topped features of the Mid-Pacific Mountains.

These hypotheses are that the flat-topped features may be: (1) Atolls: (a) submerged coral atolls whose lagoons were filled by pelagic sediments or filled by coral growth prior to submergence (Shepard, 1948, p. 260), or (b) atolls whose coral was killed by relative drop in sea level, with subsequent solution and erosion of reefs and final submergence.

(2) Submerged, uneroded volcanoes with coral on the rims of craters filled by pelagic sediments.

(3) Submerged, uneroded volcanoes (a) with *calderas* rimmed by coral and filled by pelagic sediments, or (b) with a period of emergence and erosion followed by submergence.

(4) Eroded, up-faulted portions of the Mid-Pacific Mountains upon which coral became established, and which were then submerged.

(5) Submerged truncated volcanoes with coral present as banks; smoothing of the tops due to pelagic deposition.

Several lines of evidence deny that the guyots are submerged atolls:

(1) In fully developed modern atolls calcareous organic material completely covers the top and upper side slopes of a supposed foundation. Erosional basaltic debris dredged from the top and uppermost slopes of the guyots deny that these features were fully developed atolls.

(2) Great stress has been laid on side slopes by authors (Kuenen, 1933; 1950; Emery *et al.*, 1954) when comparing atolls with volcanoes. Kuenen (1933, p. 96–98) concludes that practically all atolls are partly surrounded by slopes of more than 45° down to 200 m, and these steep slopes may go to 600 m. He concluded that composite, andesitic volcano slopes may be more than 35° in a few sections, but are usually no more than 25° down to 500 m. Emery *et al.* (1954, p. 128) reports that in the Northern Marshalls the atoll slopes average 32° for the first 250 fathoms and that the guyots average 14° in the same portion. Cinder cones slope away from the craters at about 26°–30° (Lobeck, 1939, p. 664). Lava domes rarely exceed 5°–10° on upper side slopes. Submarine slopes off Kauai, Hawaii, are no more than 16° and off Kaula the slope is 19° (Menard and Dietz, 1951, p 1282).

Hess and Cape Johnson guyots (the only ones on which coral was dredged) have side slopes near the top of less than 22° and 17°, respectively. The conclusion is that these slopes are much closer to volcanic slopes than they are to present-day atoll slopes.

Authors (Kuenen, 1933, p. 98, and others) have noted that the steep upper slopes of atolls are a reflection of upgrowth on subsiding platforms, or a reflection of the rising waters of post-glacial time. It seems probable that the steep uppermost slopes of modern atolls are due in large part to fast upgrowth during post-Pleistocene time.

A pre-Pleistocene atoll formed on a still-standing platform need not have the steep uppermost slopes of a modern atoll. If this ancient atoll were formed on a basaltic volcano of low side slope and with no relative rise in sea level it might initially (prior to full lateral development of the reefs) have very low slopes. Therefore, conclusions based on comparison of side slopes are largely invalid in discussions of the guyot problem. One should not attach too much meaning to conclusions based on a comparison of present-day atoll slopes and the slopes of Cretaceous guyots.

It is extremely unlikely that the flat tops of guyots are sediment-filled craters. No known craters are large enough to give their volcanoes a flat-topped appearance in profile (such as on the guyots) should they be filled to rim level (Emery *et al.*, 1954, p. 128).

Submerged volcanoes with calderas filled with pelagic sediments could easily explain some guyots. Many great, modern calderas are known. Cotton (1944, p. 305, 313) lists Crater Lake, Oregon, as being a caldera 5½ miles in diameter, and that of Aso, Kyusyu, Japan, as being 150 square miles in area. Williams (1941, p. 277–283) lists several great calderas of Kyusyu, Japan: Aira, 24 by 23 km; Ibusuki, 26 by 12 km; Kikai, 22 by 13 km; and the largest, Kuttyaro, 26 by 20 km. Hess Guyot, the largest of the two coral-bearing guyots is about 8 by 12 nautical miles (about 15 by 22 km) across the top. Modern calderas, therefore, are well within the magnitude of area of many guyot tops.

A large volcano with a collapse or blow-out caldera which was submerged just beneath the surface could easily have coral banks established on the caldera rim. If this volcano were later submerged rapidly enough to kill the coral and then filled with pelagic sediment it would present a flat-topped basaltic feature with coral on the edges near the break in slope. Volcanic debris would be present on the side slopes. Some of this debris might be erosional if portions of the rim had remained above water and had been eroded. In all respects save one this postulated feature would present all of the evidence previously discussed in this paper. A well-rounded olivine basalt pebble was dredged from the center of the top of Hess Guyot. In the postulated case above there could be no erosional debris in such a position. If the postulated submerged caldera had been elevated above the surface and even slightly eroded prior to eventual submergence, erosional debris could easily have been swept onto the flat top. In this event the feature would completely fulfill all requirements demanded by the evidence.

Morphology precludes the possibility that the guyots are up-faulted blocks of the Mid-Pacific Mountains. The oval tops and the concave upward slopes which gradually merge with the deep-ocean floor do not suggest fault blocks, which are normally bounded by linear or irregularly curved escarpments. Emery *et al.*, 1954, p. 128 came to the same conclusion with respect to the guyots of the Northern Marshall Islands.

The remaining possibility, that the coral-bearing Hess and Cape Johnson guyots are wave-truncated submerged volcanoes on which corals were present as banks on and among the basaltic debris, is the most simple and likely landform represented by these features.

The following evidence supports the probability that the guyots were volcanoes:

(1) The seamounts have oval plans and symmetrical profiles.

(2) The concave upward side slopes of the seamounts are less than 22° near the top and at the foot of the slopes they gradually merge into the deep-sea floor south of the main ridge of the Mid-Pacific Mountains.

(3) Basaltic volcanic debris is associated with the guyots.

(4) The guyots are similar in morphology and rock type to many known volcanic islands and seamounts. They are directly connected to the Hawaiian chain of basaltic, volcanic islands.

(5) The guyots are more or less in linear alignment.

Truncation of the guyots is deduced by the flat tops, the rounded and subrounded sand grains, pebbles, cobbles, and boulders of basalt and the coral which proves that the seamounts were once at or near the surface and close to the zone of wave abrasion.

The presence of erosional debris with coral near the break in slope on top of the features precludes the possibility that the seamounts were fully developed atolls and indicates a probability that the coral was present on and among the erosional debris.

TRUNCATION OF THE GUYOTS

The time of truncation is known, from the age of the fossil fauna which lived on the tops of the guyots, to be pre-Aptian to Cenomanian.

There are three general types of material which the waves might have abraded:

(1) Solid olivine basalt: data are not available on the rate of erosion of solid olivine basalt, but the time necessary to abrade an olivine basalt volcano must be on the order of several millions of years.

(2) A composite of pyroclastic material and lava flows: such material would be much more quickly eroded than homogeneous basalt; the pyroclastic material would be quickly eroded, which would undercut the lava flows and aid their erosion by caving and slumping.

(3) Pyroclastic debris: this material would be very quickly eroded to a flat platform; Hoffmeister *et al.* (1929) and Hoffmeister and Ladd (1944, p. 392) describe the erosion of Falcon Island in the Tonga group which was composed entirely of pyroclastic material. Falcon Island was reduced to a bank 5–10 fathoms deep between the time of its eruption in 1885 (when it was 300 feet above sea level) to 1912. Wharton (1897, p. 392) noted that Sabrina Island in the Azores was formed in 1811 by the accumulation of pyroclastic material which was soon washed away to a depth of 15 fathoms. Hoffmeister and Ladd (1944, p. 398) believe that in the coral island area of the Pacific pyroclastic islands were common and list Vitilevu, Lau Islands, Falcon, Eua, and the New Hebrides as being made up dominantly of pyroclastic material. They also believe that the platforms on which atolls were formed were largely pyroclastic cones.

There are several possibilities as to how far back in geologic time the truncation of the Mid-Pacific guyots took place. Some of these are:

(1) Truncation in Precambrian, slow submergence in a coral-free sea, with elevation in the late Mesozoic for lodgment of coral followed by submergence; or, as implied by Shepard (1948, p. 277) a drop in sea level might allow coral to lodge on deep guyots.

(2) Pile-up of the ridge of the Mid-Pacific Mountains in the Paleozoic with emergence of the upper peaks in the late Paleozoic or early Mesozoic; commencement of the planation of solid basalt with no coral lodged on the benches owing to non-migration or unknown ecologic factors; final abrasion of the flat tops in late Mesozoic followed by lodgment of coral and final submergence.

(3) Emersion of the peaks in the late Mesozoic; planation (which could have been rapid if the peaks were composite or pyroclastic cones); the lodgment of coral followed by submergence.

If the guyots were composed of solid basalt a long still-stand at the surface would have been required to complete planation of the flat tops. If this had happened in late Mesozoic the corals probably would have lodged themselves on the initially eroded flat benches and thereafter have protected the interior portions from further wave abrasion. No such interior portions were observed on the guyots. It seems probable that the islands were completely truncated prior to lodgment of the coral.

The coral could not have lived on Hess and Cape Johnson guyots for long before submergence because it did not spread and bury the basaltic erosional debris.

Information is scarce on lavas extruded under water. Macdonald (in Emery et al., 1954, p. 123; and personal communication) believes that much pyroclastic material is formed by subaqueous flows, especially near the surface, and that lavas extruded in deeper waters are apt to be less vesicular and produce less pyroclastic material. Macdonald emphasizes that the type of lava flowing into the sea markedly affects the production of cinders and pyroclastic material. If the flow is fragmented on top (as in the aa type of lava) the sea water enters the material and steam explosions result; if the flow is unfragmented (as in the pahoehoe type of lava) the water cannot enter the material and pillow lavas are more apt to form. Mathews (1947) summarized author's opinions as to the form taken by subaqueous lava flows. Sampson, Fuller, and Koto believe that the lava disintegrates into fine breccia; Anderson (in Mathews, 1947, p. 568) believes pillows are also developed. It is probable that much pyroclastic material, fine breccias, and pillows are produced. Stearns (1946a, p. 17) in postulating the geologic history of the Hawaiian Islands, believed that in the initial, submarine phase the lavas were chiefly pillows, ash, and pumice; and that when a cone first emerges from the sea it is composed largely of weakly consolidated ash which is quickly eroded by the sea; shortly afterwards lava flows veneer the cones and reduce the erosive effectiveness of wave action. These opinions coincide with those of Hoffmeister and Ladd (1944, p. 469–470). If true this sequence would explain rapid erosion of the guyots to flat platforms upon emergence in the Mesozoic.

No final answer can be given to these problems. The simplest hypothesis is that peaks of the main ridge of the Mid-Pacific Mountains appeared above the surface of the sea during late Mesozoic, that they were composite cones of lava and pyroclastic material, or entirely pyroclastic, and so were rapidly eroded to flat banks with the sediments of erosion inhibiting coral growth (as at Falcon Island in 1928), that about the Middle Cretaceous (Aptian to Cenomanian) coral became lodged on these flat banks and was soon killed by rapid submergence prior to the end of the Cenomanian.

GEOMORPHOLOGY OF THE MID-PACIFIC MOUNTAINS

The gross morphology of the Mid-Pacific Mountains is similar to that of the Hawaiian structure: a linear ridge rising from a broad swell of the ocean floor (Dietz and Menard, 1953, p. 106). Both structures are embayed and have similar low slopes which merge gradually into the deep-sea floor. The conclusion is that the structure of the Mid-Pacific Mountains is similar to the Hawaiian structure.

Chubb (1934, p. 295) postulated that the Hawaiian Islands are on an anticlinal ridge in which fissures have opened up and extruded lavas have formed the volcanoes.

Betz and Hess (1942, p. 101) noted that except for the peaks of the Hawaiian Islands the region is a very gentle swell on the deep-sea bottom. They postulated that the opening of fissures, perhaps tension cracks, or transcurrent faults on the floor of the ocean, and extrusion of large amounts of volcanic material built up the swell and islands.

Stearns (1946a, p. 16) believed that the Hawaiian Islands are a chain of shield-shaped basaltic domes built over a fissure 1600 miles long on the ocean floor. The fissure may be a tear along the crest of the fold, a simple tension crack, a group of en echelon faults, or a strike-slip fault. He believed that the lava rises along tension cracks bounding lozenge-shaped blocks strung out linearly from south-east to north-west, with vents about 25 miles apart. Fissure eruptions characterize Hawaiian volcanoes.

Dietz and Menard (1953) explain the newly defined arch and deep around the southeast end of the Hawaiian chain, and the Hawaiian structure, by postulating that the Hawaiian swell was formed by a broad uplift of the sea floor due to upwelling and thermal expansion of the hot magma of a subcrustal convection current, followed by volcanism through tension fractures at the crest of the swell, and then by regional and local isostatic sinking.

Hess (1954, p. 344-347) notes three working hypotheses for the origin of oceanic ridges: (1) extrusion of basaltic lavas and pyroclastic materials along a line of fractures in the crust (*same as* Betz and Hess, 1941), (2) brecciation of the peridotite substratum and the rise of a column of hotter and less dense material due to an upward convection current in the mantle, and (3) thickening and buckling of the upper basaltic crust and resultant volcanism because of a downward limb of a subcrustal convection current. Hess notes in his discussion of (2) that an upward convection current might help lift the ridge and cessation of the current let it subside faster than would be normally expected. This latter interesting idea was mentioned to the writer by Heiskanen (Personal communication, 1951) as an alternate explanation of rapid subsidence.

The simplest explanation for the origin of the Mid-Pacific Mountains (without regard to motivating forces) appears to be that they are a linear series of ridges and volcanoes formed by the extrusion of basalt and pyroclastic material from the crest of a low swell through tensional fissures or faults; and that the fissures or faults intersect those of the Hawaiian ridge at Necker Island so that the two ranges are connected. (*See* Pl. 10.) The fractures through which the lavas and pyroclastics of the Mid-Pacific Mountains and the Hawaiian ridge were extruded may belong to great structural trends of the northeastern Pacific (Menard, 1955, in press).

Plate 10 shows that the general similarity between the Mid-Pacific Mountains and the Hawaiian Ridge ends at about 180° W. Long. Soundings in the northwest portion (between 170° and 180° E. Long.) are inadequate. The southwest portion of the Mid-Pacific Mountains are under the main shipping routes between the Marshall Islands and the Hawaiian Islands and here the bathymetric control is good. Evidence is sufficient to suggest that the westernmost part of the Mid-Pacific Mountains is formed by basalts from a complex of parallel and intersecting fissures or faults.

It should be noted that the guyots of the Northern Marshall Islands were not formed in the same way as the guyots of the Mid-Pacific Mountains. The guyots of the Marshall Islands rise as separate volcanic cones from the deep-sea floor (Emery *et al.*, 1954). The solution of the problems posed by the guyots of the Mid-Pacific Mountains by no means solves the problems of all guyots.

The guyots of the Mid-Pacific Mountains do not have accordant tops. This is not surprising considering they are volcanic. A volcanic cone could have been eroded to a flat top and have subsided (with the whole structure) before another emerged and was eroded. These two tops would not then be on the same level. The shallowest sounding on the Mid-Pacific ridge was at MP 34 (Pl. 11) where sharp, apparently uneroded peaks came within 567 fathoms of the surface. This could be a late volcanic cone which did not reach the surface and was, therefore, uneroded.

The Mid-Pacific Mountains and their guyots are fossil landforms from the Cretaceous. They have been preserved from erosion by submergence into the deep Pacific where they exist today as extremely ancient mountains now altered only by the slow rain of pelagic sediments from above.

CAUSES OF SUBMERGENCE

The first important problem concerning causes of submergence is to determine whether the ocean level rose or the bottom sank. This involves a discussion of the cubic content, or volume, of the ocean during geologic time. This problem has been recently studied by several workers, and there is a growing body of opinion that the volume of the oceans has not materially increased since the Paleozoic. Kuenen (1937, p. 460–462; 1950, p. 129–131), in a long summary of the supporting evidence, notes that the only possible source of new waters is from magmatic juvenile waters, the total volume of which would have little effect on the volume of the seas after the Precambrian. He further notes that there is no paleontologic evidence that the seas were less saline in post-Algonkian times, which would have occurred if salts had been delivered gradually to the oceans through weathering while the water was produced principally during the last quarter of terrestrial history. He concludes that it is a reasonable postulate that the total amount of sea water has not been augmented by more than a fraction since the beginning of the Paleozoic. Umbgrove (1947, p. 235) summarizes Kuenen's observations and agrees with his conclusions.

Rubey (1951) summarized the evidence to show that the oceans came from the interior of the earth (juvenile waters) and that the volume has grown with time with a concurrent increase in growth of continental masses and a progressive sinking of ocean basins. He believed (1951, p. 1124) that the volatiles have risen to the surface gradually and not in a few great bursts.

Kulp (1951) produced an independent argument for the origin of the hydrosphere from the interior of the earth which agreed in essence with Rubey. In 1950 the writer asked Rubey if he would estimate the total per cent of ocean volume added since Middle Cretaceous time. Rubey replied that

... if the rate of increase has been constant through the past then the estimated time that has elapsed would lead one to expect only a few per cent additional water since then (Middle Cretaceous). If your Cretaceous fossils at 900 or 1000 fathoms below present sea level have been buried solely by an increased volume of sea water without any local sinking whatever they seem to imply a percentage increase of about 30 per cent, even after making due allowance for the probable volume of marine sediments that have accumulated since the Cretaceous.

The writer posed the same question to Kulp, who replied that if the rate of ocean growth from the interior is statistically linear, then the total ocean volume from Middle Cretaceous should be about 3–4 per cent, or something on the order of 50–100 meters increase in depth.

R. R. Revelle (personal communication) is currently considering various aspects of the depth of sediments in the Eastern Pacific and the drowning of the guyots and believes that a large part of the present volume of the oceans may have been produced from below the present sea bottom when the lower part of the unconsolidated sediments were metamorphosed sometime after the late Mesozoic. With an increase in the volume of the oceans the sea floor sank, thus he would explain the relative submergence of the guyots as owing mostly to rise of sea level with some sinking of the ocean floor.

Hess (1954, p. 346–347) proposes that water may be produced below the ocean floor by the process of serpentinization of the peridotite below the Mohorovičić discontinuity. He thinks that this alteration would be local, or at most, regional and would release an amount of water which would have little effect on the rise of sea level around the world.

From the previously cited papers the point appears to be well taken that the ocean waters came from the interior of the earth. The question as to the *rate* of this accretion has not yet been determined. The writer favors the opinion of Kuenen, Umbgrove, and others that this accretion since the Cretaceous has been on the order of only a few per cent; and that if the material underlying the unconsolidated sediments was strongly metamorphosed with the release of large amounts of water (as implied by Revelle) that such an event was relatively regional and the consequent amount of water released had little effect on worldwide sea level, which is the opinion of Hess (1954, p. 347).

Hess's postulate (1946, p. 790) that submergence of the guyots was due largely to relative rise of sea level from sedimentation in the ocean basins since Precambrian time is no longer tenable in view of the Cretaceous age of the guyots. Raitt (1953, p. 3) as a result of seismic work estimates total depth of unconsolidated sediments in the Eastern Pacific to be about 300 m. Oliver et al. (1955, p. 945) conclude from a study of earthquake surface waves that the thickness of the sedimentary layer in the Pacific is 0.5–1.0 km. They find a similar range in the Atlantic. Presumably only a part of this sedimentation occurred since Middle Cretaceous time, and so can account for only a small part of the submergence. Kuenen (1950, p. 398) estimated 600 m relative rise of sea level owing to sedimentation since the beginning of the Cambrian. He adds 175 m for addition of juvenile waters to get a total of 775 m.

Some submergence must have been due to compaction of soft oceanic sediments under the extruding basalt. Kuenen (1950, p. 397) attributes a possible subsidence of several hundred meters to this cause. This may be true of a single volcanic cone, but under a great basalt range of the magnitude of the Mid-Pacific Mountains this compaction would have largely occurred long before the topmost peaks emerged and were truncated. This factor would, therefore, appear to be negligible in explaining the final submergence on the order of a mile.

Isostatic adjustments undoubtedly play a part in the submergence of many oceanic islands and ridges. In recent years the ideas about such adjustments have changed significantly. Vening Meinesz (1941; 1948) has shown that the crust of the Pacific behaves as a strong and elastic solid rather than a viscous liquid. A small volcano could be supported by this crust with no isostatic compensation. As the volcano grew the crust would bend down (according to Vening Meinesz). Gunn (1943; 1947; 1949) published a series of papers on isostasy in which he proposed that a great load such as that of the Hawaiian Islands would be borne partly by the elasticly downbowing crust and partly by hydrostatic pressures of the underlying lithosphere. He termed this state of vertical mechanical equilibrium "Isobaric Equilibrium." Gunn (1949, p. 268) thinks that such adjustments occur in most cases in a relatively short geologic time span, probably on the order of about 25,000 years.

When the first basalts of the Mid-Pacific Mountains were extruded there would have been no isostatic adjustments because these flows would be supported by the strength of the crust. As the mass increased a point would be reached at which the elastic crust would begin to downbow. This downbowing would continue as the mass increased, the additional load being partly compensated for by the hydrostatic pressures of the lithosphere. By the time the topmost peaks of the range were formed the huge mass of basalt should have caused an appreciable downbowing. In other words, isostatic adjustments should have started long before the guyot peaks were truncated. Vening Meinesz (1934; 1948a) has reported on a line of gravity stations between San Francisco and Guam. In his discussion of the meaning of the gravity anomalies across the Hawaiian Islands he concluded that the data point to regional isostatic compensation distributed over a large area, *viz.*, up to a radius of 170–240 km. He believed that this made it probable that a rigid crust 25–45 km thick is present. West of Hawaii Vening Meinesz took eight stations on or near the Mid-Pacific Mountains (1948a, p. 160). Station 116 is on the deep-sea floor to the east of the Mid-Pacific Mountains; Stations 117–120 are on the main ridge; Station 121 is on the deep-sea floor to the south of the main ridge; Station 122 is near a north-south trending ridge in the southwest part; and Station 123 is on the deep-sea floor to the southwest of the Mid-Pacific Mountains.

Vening Meinesz (1948a, p. 98), in his study of the gravity anomalies over the Hawaiian Islands, determined that reduction to be nearest correct which gives the smallest value of the mean of the anomalies for any group. Thus, if the anomalies for the main ridge of the Mid-Pacific Mountains are considered as a group it can be seen (Vening Meinesz, 1948a, p. 160) that those of smallest value are for a regional radius of compensation of 116.2 km and a crustal thickness of 20 km. All of these anomalies are negative and range from −1 to −17 milligals. These small values indicate that the Mid-Pacific Mountains are essentially in isostatic equilibrium, and that, as in the Hawaiian Islands, the most logical explanation is one of isostatic compensation distributed over a large area, which means that the crust is bending downwards over this area under the weight of the great mass of submarine mountains. A possible explanation for the negative anomalies might be contained in a paper by Gunn (1949, p. 278), who points out that a gravity anomaly may exist owing to deficiency of mass under a load (this mass having been squeezed out to the sides, with the overlying weight partially borne by the elastic strength of the crust) even though the load is in perfect equilibrium.

If isostatic adjustments were far advanced (as seems probable) when the top peaks of the range emerged, then little of the eventual guyot submergence can be attributed to isostatic adjustments owing to the load of the volcanic ridge.

Most writers, in considering the increase in ocean volume in geologic time, hold that the ocean basins sank as the volume of oceanic waters increased. Thus the present position of the sunken islands must be partially due to isostatic adjustments under the volcanic ridge but also to a general sinking of the ocean floor.

Hess (1954, p. 347) advances another cause of submergence—the rise of isotherms in a serpentinized area, as might happen over the upward limb of a convection current or as a result of extensive volcanism, with a consequent reversal of the serpentinization reaction, the release of water, and subsidence of the ocean floor.

For years many geologists supposed that the Pacific Basin has been a stable mass throughout geologic time. Seismic studies by Gutenberg and Richter (1941, p. 82) reveal that the Pacific Basin is at present the largest of all stable masses. It appears to be an area of complete seismic calm with almost no epicenters inside the Andesite Line. However, evidence is rapidly accumulating which proves that the Pacific Basin has suffered subsidences, uplifts, volcanic activity, faulting, and other geologic activity.

Inasmuch as appreciable increase in ocean volume, sedimentation, compacting of soft sediments, and isostatic adjustments have been discounted as major factors which could completely explain the submergence of the guyots, probably only tectonic activity or crustal movement, together with the preceding factors, could have caused deep subsidence of the area of the Mid-Pacific Mountains. Both Schuchert (1932, p. 537–561) and Kuenen (1950, p. 549) postulate crustal movements to explain great eustatic changes. There is little evidence of large-scale faulting in the Mid-Pacific Mountains.

Griggs (1939) and Vening Meinesz (1944; 1948b) have proposed deep subcrustal convection currents to explain significant movements of the crust which cause deep subsidences. Kuenen (1950, p. 549) favors this explanation as does Heiskanen (Personal communication). The writer favors subsidence of the Pacific Basin in the area of the Mid-Pacific Mountains owing dominantly to crustal movements possibly caused by the subcrustal convection currents postulated by Griggs and Meinesz. In the writer's opinion isostatic adjustment, sedimentation, compaction, and rise of sea level are not adequate to account for the great depths to which the Mid-Pacific Mountains and their guyots have been submerged. However, if one is prepared to accept the ideas of Meinesz and Gunn that the great loads on the Pacific crust are borne partly by the elastic strength of the downbowing crust and partly by hydrostatic pressure of the underlying lithosphere, then the idea should be carried to its ultimate conclusion: under an excessively great load a point might be reached beyond which the strong crust would reach its elastic limit and the mass previously supported in part by the crust would founder and seek its isostatic level.

Whether the major part of the submergence of the guyots was due to foundering of the great load of the Mid-Pacific Mountains or to subcrustal forces such as downward-moving convection currents (or cessation of upward currents), it appears proven that a subsidence of the Pacific Basin of several hundreds of fathoms did take place in the area of the Mid-Pacific Mountains.

BEARING OF CONCLUSIONS ON CORAL-ATOLL HYPOTHESES

The findings of the present study are significant in the classic geologic controversy over the formation of coral atolls. Excellent summaries of this problem are readily available (Davis, 1928; Shepard, 1948; Vaughan, 1919; Ladd and Tracey, 1949), and it is necessary here to discuss only those ideas affected by the discoveries of the 1950 Mid-Pacific Expedition.

In general, the conclusions of this study which pertain to the coral atoll problem are as follows:

(1) The guyots of the Mid-Pacific Mountains are truncated volcanoes with flat platforms which have been eroded by wave action.

(2) Reef-building corals were established on these platforms during Cretaceous time, but were probably killed by rapid submergence.

(3) The guyots have probably been submerged by subsidence of the sea bottom to a depth of about 900 fathoms.

The evidence for subsidence in this paper is another addition to the growing body of information which suggests that subsidence of seamounts has taken place on a large scale in the Pacific Basin. (Drilling on Bikini: Ladd *et al.*, 1948; seismic survey of Bikini: Dobrin *et al.*, 1949; drilling on Kata-Daita-Zima: Hanzawa, 1940; gravimetric survey of Jaluit Atoll: Matsuyama, 1918; drilling on Funafuti: Royal Soc. London Report, Hinde, 1904; drilling on Eniwetok: Ladd *et al.*, 1953; seismic surveys in the Marshall Islands: Raitt, 1952).

In 1837 and 1842 Darwin published his classical papers on the formation of coral reefs and atolls. His principal postulate was the well-known theory that barrier reefs and atolls develop from subsidence of fringing reefs. The idea was widely accepted and Darwin became the "father" of the Subsidence Theory. In the welter of subsequent controversy the fact that Darwin suggested other possible ways in which atolls could form was largely overlooked. Vaughan (1919, p. 242) and Davis (1928, p. 35) mention this, but most authors disregard the logical and temperate auxiliary hypotheses of Darwin, such as these quoted from the third edition of his 1842 book (1889):

I remarked—that a reef, growing on a detached bank would tend to assume an atoll like structure; if, therefore, corals were to grow up from a bank some fathoms, submerged in a deep sea, having steep sides and a level surface, a reef not to be distinguished from an atoll might be formed; and I believe some such exist in the West Indies. (p. 121)

If we consider, moreover, the number of atolls in the midst of the Pacific and Indian Oceans, this assumption of so many submerged banks is in itself very improbable. (p. 122)

I may here observe that a bank, either of rock or of hardened sediment, level with the surface of the sea and fringed with living coral, would be immediately converted by subsidence into an atoll without passing, as in the case of a reef fringing the shore of an island, through the intermediate form of a barrier reef. *As before remarked if such a bank lay a few fathoms submerged the simple growth of the coral, without the aid of subsidence, would produce a structure scarcely to be distinguished from a true atoll.* (p. 138) [italics by the writer].

Darwin discounted (p. 138–139) this explanation for "the larger groups of atolls in the Pacific and Indian Oceans".

Darwin first postulates the idea that some atolls could have been formed on flat, subsiding platforms and have grown into atolls without having passed through a barrier-reef stage. This is now generally recognized as an extremely probable mode of formation of some atolls of the Marshall Islands which appear to have been formed on guyots (Eniwetok and possibly Bikini). Darwin's work anticipates by many years another theory of atoll formation, that of the "antecedent platform" proposed by Wharton (1897) and Hoffmeister and Ladd (1944). They claim no originality but neglect to refer to Darwin's original statement. Darwin should be credited with the original idea even though he discounted it in favor of his more widely known hypothesis. He stated in clear words that an atoll could form by simple growth of coral from a still-standing level platform.

Darwin further remarked (p. 146):

I have shown that there sometimes exist in the neighborhood of atolls, deeply submerged banks with level surfaces; that there are others less deeply but yet wholly submerged having all the characters of a perfect atoll, but consisting merely of dead coral rock—these several cases are, I believe, intimately related and can all be explained by the same agency of subsidence.

Exploration has remarkably borne out these statements. Bikini Atoll with its adjacent guyot (Sylvania Seamount) is a type example of the first statement and many examples of drowned atolls are listed by Davis (1928).

Daly in 1910 proposed his Glacial Control theory and continues to defend it. Although his original idea has been modified and restricted, most authors admit that glacial control has had important modifying effects on present-day reefs. Daly erred in discounting subsidence as a major factor in atoll formation, a subsidence now proved to be general in the coral-reef area of the Northwestern Pacific and in the area of the Mid-Pacific Mountains.

An important contribution was made to the coral-atoll problem by Hoffmeister et al. (1929) and Hoffmeister and Ladd (1944) when they called attention to the extremely widespread and quantitatively important presence of pyroclastic material in some of the Pacific islands. They pointed out that a volcanic form composed of pyroclastic material will be quickly eroded to a submerged bank. The belief that an extremely long time was required to truncate a volcanic form was one reason why Davis (1919) discarded the platform hypothesis. Davis (1919, p. 433) admits:

however, it is, I believe, possible that a still-standing volcanic island should, after its eruptive construction, be cut away by ocean waves, because no reefs can be formed on the beach of loose detritus that would surround such an island while abrasion is in progress.

Davis also discounted the platform theories because he thought it unlikely that a volcanic form which was built up and later subsided would stand still long enough for a flat platform to be eroded. This would be logical except in the case of a volcanic form largely composed of pyroclastic material.

The Mid-Pacific guyots furnish an example of coral lodgment on a flat, antecedent platform. If submergence in the area had been slower the Cretaceous coral would have remained at or near the surface and would have probably formed an atoll. The eventual results might well have been similar to Bikini Atoll where a bore hole 2556 feet deep has penetrated Lower Miocene reef material and the underlying foundation is thought to be more than 4200 feet deep (Raitt, 1952, p. 21) or to Eniwetok Atoll where the foundation was determined by drilling to be 4154 feet deep on the southeast side of the atoll where Eocene material overlies olivine basalt (Ladd et al., 1953, p. 2257).

The results of Mid-Pac are considered as proof of subsidence and, therefore, strongly support Darwin's subsidence theory with its auxiliary hypotheses and Davis' (1928) able defense of Darwin's ideas; they affirm the presence of flat, eroded, antecedent platforms on which coral had found lodgment and might well have grown into an atoll, proving that Cretaceous reef coral was present on platforms in the Pacific and possibly forecasting an age for the incipient reefs of some of the Pacific atolls.

Given the favorable ecologic factors of reef-coral growth, including a foundation whose top was within the depth zone of reef growth, a coral reef will grow to the surface regardless of still-stand or submergence (granted slow subsidence or submergence). It is theoretically possible to get all the various kinds of reefs with and without submergence. No theory is all-embracing, and attempts to explain all reefs with one theory are destined to failure. The happy conclusion here is that almost everyone is right at some time and in some places (as pointed out by Stearns, 1946b, p. 262).

MODIFICATION OF THEORIES OF FAUNAL MIGRATION

The anomalous distribution of some plants and animals has long intrigued biologists and geologists. The close affinity, or identity, of faunas or individual animals on opposite sides of the Atlantic and Pacific oceans has been the subject of a great body of literature. Few authors (including the present writer!) can resist the temptation to speculate on the methods by which the animals migrated across the oceans. In general, there are three explanations offered by students of animal migration (Davies, 1934, p. 44, 45). The first, that migration can be explained by Continental Drift (Wegener, 1922, and others), is not pertinent to problems in the Pacific since the postulates of the theory of Continental Drift do not call for juxtaposition of continents in that area. The second explanation, that the oceans and continents are permanent features of the earth and that migrations can be explained by presently indicated land bridges (Bering, Central American, etc.) or by overseas migration, has been championed by Wallace (1880), Matthew (1915), Simpson (1940), Clark (1945), and others.

The third explanation, that faunal migrations can best be explained by trans-oceanic ancient land connections or continents now sunken, has the support of many writers (Gregory, 1930; Campbell, 1919; Pilsbry, 1899; Cooke, 1923; Skottsberg, 1926; and others).

The anomalous distributions refer in the main to stenothermic shallow-water animals (those which tolerate a short temperature range). These animals can migrate freely in the shallow waters along continental shelves within their temperature range, but cold waters are barriers to their dispersal around the northern and southern seas. It is known, however, that identical species of shallow-water marine invertebrates of the tropical, semitropical, and temperate zones are found on opposite sides of the oceans at present and in the geologic past. Hertlein (1937, p. 305–309) lists 29 species of shallow-water tropical and semitropical mollusks and echinoids which are found at present along the shores of North and South America and in the Polynesian Islands. The rudistids and corals are examples from the past.

Gregory (1930) presented a long summary of the evidence for ancient continents in the Pacific to explain plant and animal migrations. Gregory concludes that the evidence that floras and faunas of Hawaii and other oceanic islands of the Pacific are allied to those of the Indo-Pacific can best be explained by a great continent across the Central Pacific which persisted into the Cenozoic. He quotes the opinions of many other writers (some listed previously) to the same effect.

The presence of ancient Cretaceous islands in the deep Cretaceous sea of the area of the Mid-Pacific Mountains destroys at one blow the concept of a continent in the area in the Cretaceous and Cenozoic eras, and with this concept a whole body of literature explaining faunal migration across this ancient, sunken continent.

The discovery of Cretaceous islands in the Mid-Pacific is one of those agreeable findings which destroy one concept only to present another in its place. The discovery does away with the idea of a Cretaceous or later Pacific continent and at the same time presents interesting possibilities for explaining faunal migrations by "island stepping stones."

There are several ways in which shallow water animals can migrate over deep waters. These are: (1) by attachment to, or lodgment on floating material such as logs, natural rafts, and pumice, (2) by transportation of planktonic larva of sessile benthonic animals in currents during the meroplanktonic stage, (3) on the feet of, or attached to fish, birds, or marine mammals, and (4) by strong winds. These several possibilities have been discussed by many authors (Lull, 1929; Davies, 1934; Clark, 1945; Vaughan and Wells, 1943; and others).

The chief objection to over-water migrations of shallow-water animals has been the tremendous distances these animals must travel over deep water before finding shallow water of the right temperature in which to find lodgment.

Ancient islands in the deep ocean would have offered intermediate stops for shallow-water animals advancing across great expanses of ocean. This is especially true of the best possibility for migration: transportation by currents during the larval stage.

Davis (1928, p. 55), in a penetrating analysis of possible proofs of subsidence, postulated sunken islands as an explanation for anomalous faunal distributions in the Pacific. Freeman (1951, p. 34) a geographer, in a discussion of "sunken continents" in the Pacific to explain faunal migration remarked:

Many geologists, however, are skeptical of past land connections for oceanic islands that rise from great depth and are remote from continents. In this connection it should be remembered that high islands have disappeared by erosion or been reduced to little atolls; and, hence, in the past organisms might not have had to migrate as long distances across the ocean as a cursory examination of the present map would seem to indicate as necessary.

The planula larva of coral exists in the meroplanktonic stage from 1 to 30 days (Vaughan and Wells, 1943, p. 68). During this period the normally sessile (or attached) coral is distributed by marine currents. The currents of the Pacific are known to attain a velocity of 2 knots (Sverdrup et al., 1946, p. 709). During the larval stage, therefore, the coral might travel a maximum of over 1400 nautical miles in oceanic currents. The common American oyster (*Ostrea virginica*) exists in a meroplanktonic stage for about 2 weeks and might travel over 600 miles in a 2-knot current.

During the Cretaceous period the Tethys Sea lay across Mexico and Central America, and there was undoubtedly an almost world-encircling east to west current at that latitude. The guyots of the Mid-Pacific Mountains lie about due west of southern Mexico, therefore it is not surprising that the affinities of the seamount faunas are with the Caribbean-Mexico faunas.

Flat-topped seamounts which were probably ancient islands are now known to be extensively distributed over the Northwestern Pacific from about the longitude of the Hawaiian Islands. Several are known between Hawaii and the mainland of North America (Ranger Bank: Emery, 1948; Erben and Fieberling guyots; Carsola and Dietz, 1952). These few guyots east of Hawaii may have been too far north to have served as Cretaceous island stepping stones, but their recent discovery and the fact that the Pacific bottom west of Mexico is little-explored serves notice that future exploration may well reveal more guyots between Mexico and Hawaii.

It is considered proved that coral and shallow-water pelecypods did migrate across the deep Cretaceous Pacific and found lodgment on the Mid-Pacific ancient islands. Thus, the fact of oceanic island migration appears to be proved. This migration may have been by attachment to floating objects or by current transportation of larva. The latter possibility is strengthened by the shorter distances now shown to be possible.

The affinity of the guyot faunas with the Caribbean area to the east is the first affinity of the Oceanic Islands to be demonstrated other than the present-day Indo-Pacific affinity. Within the area of the Western Pacific, however, the guyots furnish possibilities of inter-island migration to explain present-day faunal affinities (*i.e.* that of Hawaii with floras and faunas to the south and west).

At present the North Equatorial current passes over the Mid-Pacific Mountain area (about 17°–19° N. Lat.) from east to west. Just to the south (between 3°–10° N. Lat.) the Equatorial Counter Current moves west to east. Hurricanes in the area move, in general, from south to north.

At present the movement of the Equatorial Counter Current to the north in the summer causes some areas to have alternating currents moving in the opposite direction. Thus, a mechanism is available to explain west to east, east to west, and south to north migrations between closely spaced ancient islands now sunk beneath the surface, which might help explain present-day Indo-Pacific faunal affinities.

SUMMARY AND CONCLUSIONS

Findings and conclusions may be summarized as follows:

(1) There is in the northwest Pacific a great, undersea mountain range with sharp and flat-topped peaks now at depths between 500 and 1100 fathoms and having an east-west lineation.

(2) The flat-topped seamounts (*guyots* of Hess, 1946) are ridges and oval, symmetrical seamounts with concave upward slopes less than 22° near the top.

(3) Two of the guyots yielded a well-integrated fauna of reef corals, rudistids, stromatoporoids, and mollusks which lived in the Cretaceous period from Aptian to Cenomanian. The fauna has affinities with those of the Cretaceous Tethys Sea to the east; it lived in marine waters of normal salinity, at depths not exceeding 50–85 fathoms, in clear waters free of sediments; water temperature was not less than about 65°F. The paleoclimate was tropical to semitropical.

(4) The megafossil fauna indicates that:

(a) The Tethyan Faunal Province extends several thousand miles into the Middle Pacific.

(b) The guyots were once at or near the surface of the Pacific and that once there was an ancient chain of islands where the Mid-Pacific Mountains now lie; the Pacific Basin has not always been a stable area.

(c) The Cretaceous Pacific Ocean was at least as deep as the present height of Hess and Cape Johnson guyots (about 1700 fathoms).

(d) The rocks of the guyots are Cretaceous, the oldest rocks of the Pacific Basin yet recorded.

(e) The speculations of Hess (1946) that the guyots were probably free of coral and were Precambrian islands are disproved in part.

(f) Previously unknown Pre-Tertiary reef corals existed in the Pacific.

(5) Microfossils date the original deposition of rounded basaltic gravel near Guyot 20171 as Upper Cretaceous. The presence of Paleocene Foraminifera on the tops of three of the seamounts independently dates these tops as Pre-Paleocene or Pre-Tertiary. The planktonic Foraminifera are shown to be allied to the Tethyan Province faunas to the east in Mexico, the Gulf Coast, and adjacent regions.

(6) The igneous rocks of the guyots are of the Hawaiian olivine basalt suite and are present as rounded to angular sand grains, pebbles, cobbles, boulders, and slabs of basalt with rounded corners.

(a) This erosional debris was dredged near the break in slope and on top of the guyots. At one station coral was dredged with sandstone.

(b) This erosional debris indicates that the peaks were basaltic features eroded by waves, that these features were not fully developed atolls whose tops were sheathed with organic calcareous material, and that the corals were probably present as banks on and among the erosional debris.

(c) The gravel of core MP 27-2 (in a basin below Guyot 20171) was probably transported by turbidity currents.

(7) The presence of Eocene Foraminifera with Recent species at the surface on one guyot, and the flat tops, indicate that the tops of the guyots are essentially areas

of nonaccumulation of fine, pelagic sediment, although globigerina ooze is present on top of all of the guyots.

(8) Silicified and phosphatized limestone caused by metasomatic replacement is unusual in deep-sea areas.

(9) The association of volcanic material with the deposition of manganese oxide confirms previous reports.

(10) Geomorphic and other evidence indicate that the guyots are volcanoes and volcanic ridges (truncated by wave action) on a great undersea range which is genetically related and similar to the Hawaiian Ridge, that the range probably formed owing to eruption of basalt and pyroclastics from linear fissures or faults in the ocean floor.

(11) In connection with the submergence:

(a) The Cretaceous seas were about the same volume as they are today.

(b) Compaction of soft, oceanic sediments under the extruded basalt took place, but was not a major factor in final settling or submergence.

(c) Sedimentation on the ocean bottom explains only a small part of the final submergence.

(d) Isostatic adjustments probably resulted in downbowing of the strong, elastic sub-Pacific crust. The total load of the Mid-Pacific Mountains was probably supported both by the downbowing crust and by the hydrostatic pressures of the subcrustal lithosphere. The writer concludes that these isostatic adjustments took place for the most part before the topmost peaks of the range emerged and were truncated and, therefore, were not a major factor in the final submergence, unless the load became so great that the whole range broke through the crust and foundered.

(e) The favored theory is that the final submergence was due to subcrustal forces such as the convection currents postulated by Vening Meinesz (1948) and Griggs (1939).

(12) In connection with atoll formation, results:

(a) Show proof of subsidence in the Pacific and support Darwin's Subsidence Theory for coral atolls with its auxiliary postulates that an atoll can form on a flat, subsiding platform or that atolls may form on flat banks near the surface *without subsidence*.

(b) Indicate the presence of flat, eroded, antecedent platforms on which coral found lodgment and might well have grown into an atoll.

(c) Prove that Cretaceous reef coral was present on platforms in the Pacific and possibly forecast an age for the first reefs on the foundations of some Pacific atolls.

(13) The presence of Cretaceous islands in a deep Cretaceous Pacific eliminates the possibility of sunken continents to explain faunal migration, but at the same time presents possibilities of explaining transoceanic migrations of shallow-water stenothermic animals by "island stepping stones" that would facilitate transportation of meroplankton and of animals attached to floating material.

GEOLOGIC HISTORY OF THE MID-PACIFIC GUYOTS

The following is the postulated geologic history of the guyots of the Mid-Pacific Mountains:

(1) Hawaiian-type basalts and fragmental material were extruded from fissures or faults in the ocean floor to form the ridges and volcanoes of the Mid-Pacific Mountains.
(2) Ridges were formed, and high peaks emerged from the sea and became islands.
   (a) The submarine portions of the ridges and peaks are probably largely composed of pyroclastic material and basalt, although they may be solid olivine basalt.
   (b) Compaction of soft bottom sediments and isostatic adjustments had largely taken place before the highest peaks emerged.
(3) Because normal downward adjustments had largely taken place, the peaks and ridges remained at a standstill long enough for wave erosion to truncate the tops to provide for lodgment and growth of the fauna.
   (a) If the emerged peaks were dominantly fragmented material mixed with basalt the truncation could have been rapid; solid basalt would require a longer period of stand-still and erosion.
   (b) A late Mesozoic (pre-Aptian-Cenomanian) truncation appears most likely.
(4) During the Cretaceous a well-integrated reef coral-rudistid fauna of Aptian to Cenomanian age found lodgment on the flat tops and began to form banks among the erosional debris. The fauna, which indicates an environment of tropical to semitropical, warm, shallow water, probably migrated in east to west currents. Growth of the reef organisms progressed far enough to form banks, but not far enough to conceal the erosional debris and form an atoll.
(5) Subsidence to a depth below 50–85 fathoms took place rapidly enough to kill the coral and other shallow-water animals. In view of the probability that the Cretaceous seas were about the same volume as they are today, and because an absolute rise of sea level caused by increase in sedimentation is unlikely, the subsidence was probably due to downward movement of the earth's crust below the Mid-Pacific Mountains.
(6) Subsequent subsidence of the Mid-Pacific Mountains took place to the present depth where they remain today as the oldest uneroded mountains known on earth. They are fossil landforms preserved in the depths of the sea where they are disturbed only by light currents and the slow rain of pelagic material from the waters above.

APPENDIX A. SYSTEMATIC PALEONTOLOGY

PART 1: MEGAFOSSILS

All specimens dredged from the guyots of the Mid-Pacific Mountains and herein described are at present located at the U. S. Navy Electronics Laboratory, San Diego, California. The type repository of the new species, and all of the remainder of the specimens will be the U. S. National Museum, Washington, D. C.

Phylum COELENTERATA

Class ANTHOZOA

Subclass HEXACORALLIA

Order SCLERACTINIA Bourne, 1900

Suborder ASTROCOENIIDAE Vaughan and Wells, 1943

Family ASTROCOENIIDAE Koby, 1890

Subfamily ASTROCOENIINAE Felix, 1898

Genus *Astrocoenia* Milne Edwards and Haime, 1848

Astrocoenia revellei Hamilton, n. sp.

(Plate 5, figure 3)

Description.—Corallum cerioid and massive, forming large hemispherical head. Corallites hexagonal, closely appressed and united by thick common wall. Calices small, shallow, depressed. Septa (24) arranged in three cycles; primary and secondary septa (six each) produced to center where they unite with well-developed styliform columella. Septa of third cycle extend one-third or less into calice.

Dimensions.—Diameter of calices about 2.1 mm. Corallum cone-shaped, about 100 mm high and 105 by 65 mm wide at top.

Type locality.—Hess Guyot (MP 33K). One specimen available for study; poorly preserved and with thin coat of manganese dioxide over top and three sides. Specimen probably detrital, having rolled down slope from growing position above. The dredge apparently broke the specimen loose from a mass of similar material.

Geologic age.—Association with the remainder of ths fossil fauna implies Cretaceous (Aptian to Cenomanian) age.

Discussion.—This species is most closely related to *A. guadalupae* Roemer from the Edwards limestone (Middle to Upper Albian) of Texas, but has more regularly hexagonal calices and distinctly larger columella.

Remarks.—This specimen was flown to Washington, D. C., by Dr. Roger Revelle upon return of the expedition to the United States and was submitted to Dr. John W. Wells of Cornell University for identification. Dr. Wells replied by telegram that "coral specimen apparently new species of *Astrocoenia* genus not known later than Oligocene in Pacific area depth probably less than 100 fathoms." The range of the genus is Triassic to Recent (one species now living in the Straits of Florida).

Astrocoenia was a typical coral genus of the Tethys area during the Jurassic and Cretaceous.

This species was named after Dr. Roger R. Revelle, Director of Scripps Institution of Oceanography.

Astrocoenia dietzi Hamilton, n. sp.

(Plate 5, figure 2)

Description.—Corallum massive. Corallites closely appressed in some portions, but usually separated by dense peritheca. Calices almost completely obliterated but probably shallow and depressed.

Septa 20 in number; first two cycles of five each extend to well-developed styliform columella. Third cycle rudimentary.

Dimensions.—Calices 1.5–1.75 mm in diameter.

Type locality.—Cape Johnson Guyot (MP 37A). One specimen available for study; poorly preserved as an irregular fragment completely coated with manganese dioxide; dredged from below break in slope, probably rolled down slope from living position above.

Geologic age.—Association with total fauna suggests Cretaceous (Aptian to Cenomanian) age.

Discussion.—This species is most closely allied to *A. meinzeri* Vaughan of the Tertiary of Cuba, but has much smaller calices and difference in number of septa.

This species was named after Dr. Robert S. Dietz of the U. S. Navy Electronics Laboratory.

Family CALAMOPHYLLIIDAE Vaughan and Wells, 1943

Genus *Brachyseris* Alloiteau, 1947

Brachyseris ALLOITEAU, 1947. An. Esc. Peritos Agric., Barcelona, Vol. VI, pp. 210–213.

Type species (original designation): *Latomeandra morchella* Reuss, 1854. Upper Cretaceous (Turonian), Gosau.

Generic description: Colonial, multiplication by intracalicular budding; exhibiting mono- to tricentric series; collines continuous, tectiform (colline with an acute summitlike ridge); pseudotheca synapticulothecal; no ambulacra; septa perforate over all their extent; simple trabeculae; simple synapticulae disposed very regularly on each trabecular segment; columella parietal, feeble.

Brachyseris wellsi Hamilton, n. sp.

(Plate 6, figure 1)

Description.—Corallum colonial and massive. Corallites separated by collines. Septa confluent between corallites, exsert; no ambulacra; septa fenestrate and connected by well developed synapticulae; septa 15 to 20, not all extending to columella; columella parietal, spongy.

Dimensions.—Diameter corallites: 1–2 mm; distance between adjacent corallites: 3–4 mm.

Locality.—Hess Guyot (MP 33 K). Many specimens available for study. This species was the most common in the dredge haul at this station. Specimens frequently coated with manganese dioxide or occur in detrital limestone.

Geologic age.—The genus *Brachyseris* is confined to the Cretaceous; the generic range is probably Hauterivian to Turonian. Association with the other fossils suggests Aptian to Cenomanian age.

Discussion and remarks.—Dr. John W. Wells identified the specimens as belonging to the genus *Brachyseris* and thinks that the species is close to *B. felixi* de Angelis d'Ossat, 1905, of the Lower Cretaceous (Hauterivian). The affinity of this species with *B. felixi* supports an Early Cretaceous age, which is in accord with other evidence discussed under "Paleontology".

This species was named after Dr. John W. Wells of Cornell University.

Brachyseris montemarina Hamilton, n. sp.

(Plate 6, figure 2)

Description.—Corallum meandroid; specimen a small (20 x 22 mm) head. Corallites mono- to dicentric and spaced 3–4 mm apart. Septa strongly beaded and branching, fenestrate, exsert. Corallites deep and separated by acute collines. Columella deep in the calice and parietal (?).

Locality.—Cape Johnson Guyot (MP 37 C). One specimen available for study.

Geologic age.—By association with the rest of the fauna at this station the age is Cretaceous (Aptian to Cenomanian).

Discussion and remarks.—Dr. John W. Wells identified this specimen as *Brachyseris* Alloiteau. It differs from *B. wellsi* in its strongly beaded and branching septa, deeper calices, and more meandroid collines.

APPENDIX A. SYSTEMATIC PALEONTOLOGY

Suborder FAVIIDA Vaughan and Wells, 1943

FAMILY FAVIIDAE Gregory, 1900

Subfamily MONTASTREINAE Vaughan and Wells, 1943

Genus *Montastrea* de Blainville, 1830

Montastrea menardi Hamilton, n. sp.

(Plate 5, figure 1)

Description.—Corallum small, massive, probably encrusting, plocoid. Corallities cylindrical and not united by costae. Calices circular, but otherwise unknown. Septa in three cycles; first two extend to columella; third joins the second before reaching columella. Septa 24 in number. Septa markedly granulated laterally. Columella poorly developed; appear as a few processes between ends of septa. Costae correspond to all septa, but not confluent between corallites.

Dimensions.—Diameter of calices about 3 mm. Distance between adjacent calices about 1 mm. Largest corallum: 30 x 30 x 45 mm.

Type locality.—Hess and Cape Johnson Guyots (MP 33K and 37A). Two specimens available for study; one from each of the localities named. Both specimens in poor preservation and covered with thin coat of manganese dioxide. Specimens dredged from below break in slope and probably rolled there from growing position above.

Geologic age.—Association with total fauna indicates Cretaceous (Aptian to Cenomanian) age.

Discussion.—This new species is closely allied to *M. pecosensis* Wells of the Early Cenomanian of Texas, but has larger calices.

Remarks.—This species is known from both the Hess and Cape Johnson Guyots—further evidence of the close faunal affinity between these seamounts.

The total range of the genus *Montastrea* is from Upper Jurassic to Recent. It is well known in the Cretaceous of Texas and adjacent regions. The genus is living today in the West Indies, the Gulf of Guinea and Brazil (Vaughan and Wells, 1943, p. 173).

This species was named after Dr. Henry W. Menard, Jr., of the U. S. Navy Electronics Laboratory.

Suborder FUNGIIDA Duncan, 1884

Superfamily PORITOIDAE Vaughan and Wells, 1943

Family MICROSOLENIDAE Koby, 1890

Genus *Microsolena* Lamouroux, 1821

Microsolena sp.

(Plate 6, figure 4)

Description.—Corallum massive. Corallites without well-defined walls. Calices unknown, but would have been scattered. Septa vary in number, but about 50 usually present. Septa extremely porous and fenestrate with abundant synapticulae, like a porous meshwork. About one half of septa appear to extend to the rather large, spongy columella. The septa are directly confluent between corallites and are straight, curved, and irregularly curved.

Dimensions.—Distance between centers of corallites 9–12 mm. Thickness of septa about 0.1 mm. Number of synapticulae along 0.5 mm length of septa average 3 or 4. Corallum of largest specimen: 32 x 35 x 20 mm.

Locality.—Hess Guyot (MP 33K). Three specimens available for study; poorly preserved as probably detrital material covered with manganese dioxide. Material was dredged from below break in slope.

Geologic age.—Association with the other fauna indicates Cretaceous (Aptian to Cenomanian) age.

Remarks.—Dr. John W. Wells identified the genus as *Microsolena* or *Polyphylloseris*; the latter differing from *Microsolena* in that the calices are raised on bosses. The surface characters of the specimen have not been preserved, and the author assigns the specimen to *Microsolena* the more numerous and widespread genus. Both *Microsolena* and *Polyphylloseris* range from Middle Jurassic to Cretaceous.

The total range of the genus *Microsolena* is Middle Jurassic to Cretaceous. Hackemesser (1936, p. 46) notes *Microsaraea distefanoi* Prevor (*Microsaraea* Koby, 1889) from the Cenomanian of Italy; Hackemesser believes that the species could belong to *Microsolena*. Vaughan and Wells, 1943, p. 148, think that *Microsaraea* Koby is a synonym of *Microsolena*, therefore the upper limit of the geologic range of that genus appears to be Cenomanian.

Dr. Wells (personal communication) notes an extremely poorly preserved specimen of *Cyathophora* on one side of one specimen of *Microsolena*. This genus ranges from Jurassic to Cretaceous, and has been noted in the Turonian of France and the Cenomanian of England and Italy.

Suborder FUNGIIDA Duncan, 1884

Superfamily FUNGIOIDAE Vaughan and Wells, 1943

Family AGATHIPHYLLIIDAE Vaughan and Wells, 1943

Genus *Diploastrea* Matthai, 1941

Diploastrea sp.

(Plate 6, figure 5)

Description.—Corallum massive, large. Corallites subcylindrical, bounded by synapticular walls. Septa (around 30) in three cycles, regularly alternating in size, larger septa extending to columella; smaller septa joined to larger about halfway to columella. Septa-costae may be confluent between corallites or meet at distinct angle. Columella small and trabecular and appears fused to inner ends of septa. Dissepiments numerous. Synapticulae found only near walls.

Dimensions.—Distance between centers of corallites 6–11 mm. Diameter of corallites 6–8 mm. Corallum was a large amorphous-appearing limestone fragment which revealed its coral structure upon sectioning. Originally the specimen was about 100 mm high by 80 x 120 mm wide.

Locality.—Hess Guyot (MP 33K). One large, poorly preserved specimen available for study; probably rolled down slope from growing position above.

Geologic age.—Association with the other fossil fauna indicates Cretaceous (Aptian to Cenomanian) age.

Discussion.—This species is closest to *D. harrisi* Wells of the Lower Cretaceous of Texas, but has much larger and less-circular corallites.

Remarks.—This specimen is extremely poorly preserved and does not show all the generic characters. Dr. John W. Wells, however, believes that it is almost certainly this Cretaceous to Recent genus.

There are a few species of *Diploastrea* older than Tertiary. Prior to 1932 the only definite occurrences were from the Oligocene of the West Indies and from living coral reefs of the Indo-Pacific. The lowermost occurrence of *Diploastrea* is from the Lower Glen Rose (Aptian) of Texas. The range of the genus is, therefore, Lower Cretaceous (Aptian) to Recent.

Suborder CARYOPHYLLIIDA Vaughan and Wells, 1943

Superfamily CARYOPHYLLIOIDAE Vaughan and Wells, 1943

Family CARYOPHYLLIIDAE Gray, 1847

Subfamily CARYOPHYLLIINAE Milne Edwards and Haime, 1857

Genus *Lophosmilia* Milne Edwards and Haime, 1848

Lophosmilia fundimaritima Hamilton, n. sp.

(Plate 5, figure 5)

Description.—Corallum a single corallite, solitary, turbinate, probably fixed. Costae corresponding to septa obscure at top and absent from bottom. Calice oblong with a deep, long fossa in which the columella was probably present as a thin lamellar sheet. Septa (24) in three cycles; first two cycles are thick, solid, and wedge-shaped and most extend to columella. Third-cycle septa are smaller, but almost reach columella. All septa have distinct vertical ridges or carinae along the sides. Septa exsert and top edges curve downward toward center. No synapticulae or dissepiments.

Dimensions.—Top: 15 mm by 11 mm. Height: about 13 mm. Width of thicker septa near wall about 0.7 mm.

Type locality.—Cape Johnson Guyot (MP 37A). One specimen of medium preservation available for study; embedded in matrix of partially phosphatized limestone in association with fragments of pelecypod shells and on base of manganese dioxide. Specimen may have lived atop this fragment of debris below break in slope.

Geologic age.—Association with the other fossil fauna indicates Cretaceous (Aptian to Cenomanian) age.

Discussion.—Species is most like *L. simplex* of the Cenomanian of France, but is larger, thicker, and has more wedgelike septa.

Remarks.—The range of this genus is definitely Cenomanian to Recent with one species reported from the Jurassic of India. The bathymetric range of this solitary coral is 100–200 fathoms in present seas (Vaughan and Wells, 1943).

Suborder FAVIIDA Vaughan and Wells, 1943

Family OCULINIDAE Gray, 1847

Subfamily OCULININAE Vaughan and Wells, 1943

(Plate 9, figure 2)

Remarks.—The dentate septal margins, the laminar, nonporous septa, absence of synapticulae, the dense peritheca, dendroid growth and well developed columella composed of twisted trabecular processes places this specimen in the subfamily Oculininae Vaughan and Wells, 1943. The poor preservation of the specimen prohibits a generic identification, but this ahermatypic coral seems closest to *Astrhelia* Milne Edwards and Haime, 1849, from the Miocene of Maryland.

Type locality.—MP 34. One poorly preserved specimen and several fragments available for study. This species was taken by dredge from among rather sharp, uneroded peaks and was probably an ahermatypic, or nonreef-building coral, which may have lived at any time within its geologic range.

Phylum COELENTERATA

Class STROMATOPOROIDEA

Order STROMATOPOROIDEA Nicholson and Murie

Family ACTINOSTROMIDAE Nicholson, 1886

Genus *Actinostroma* Nicholson, 1886

Actinostroma pacifica Hamilton, n. sp.

(Plate 7, figure 4)

Description.—Coenosteum massive, subglobular; base of attachment small. Concentric laminae and latilaminae undulating. Surface granulated by projecting ends of radial pillars; astrorhizae absent. Radial pillars hollow and continuous through many laminae and interlaminar spaces. Radial pillars less numerous and less regularly developed than laminae.

Dimensions.—Vertical elements (radial pillars) 0.32 mm wide; average number counted in 2 mm: 2–3. Horizontal elements (laminae) average 0.32–0.37 mm thick and average 5–6 every 2 mm.

Type locality.—Cape Johnson Guyot (MP 37C): one specimen. Hess Guyot (MP 33K): one specimen. Specimens well preserved in matrix of fossiliferous limestone covered by a coating of manganese dioxide. For associated fauna see remarks under appropriate stations in the section Paleontology.

Geologic age.—Association with the other fossil fauna indicates Cretaceous (Aptian to Cenomanian) age.

Discussion.—*Actinostroma* is distinguished from *Actinostromaria* by the absence (or less development) of the astrorhizae and by the less developed vertical elements. This species appears to be most closely related to *Actinostroma tokadiensis* Yabe and Sugiyama, but its horizontal elements are thicker and the vertical elements wider and fewer.

Remarks.—The geologic range of this genus is from Cambrian to Cretaceous (Senonian). Yabe and Sugiyama (1935, p. 153) note eight species from the Mesozoic.

Family STROMATOPOROIDAE Nicholson, 1886

Genus *Stromatopora* Goldfuss, 1826

Stromatopora sp.

(Plate 7, figure 2)

Description.—Coenosteum calcareous, spheroidal, made up of concentric, relatively wide laminae with vertical elements between laminae, but not extending through more than one lamina. Reticulate in vertical section and vermicular in horizontal section. Surface of laminae almost smooth with mamelons indistinctly developed. Astrorhizae absent or unknown. Horizontal elements more prominent than vertical elements.

Dimensions.—Vertical elements about 0.21 mm wide and about 7 in 2 mm. Horizontal elements about 0.26 mm thick and about 0.05 mm apart. Coenosteum of the most perfect specimen 9 mm in diameter.

Locality.—Hess Guyot (MP 33K). Three small specimens available for study; well preserved in limestone matrix. For associated fauna see the remarks under the appropriate station in the section Paleontology.

Geologic age.—Association with the other fauna suggests Cretaceous (Aptian to Cenomanian) age.

Discussion.—This species seems most closely related to *Stromatopora memoria-naumanni* Yabe and Sugiyama from the Jurassic of Japan, but has much narrower interspaces between the horizontal elements.

Remarks.—Dehorne (1920, p. 80, 81) gives the range of this genus from Silurian to Cretaceous (Cenomanian).

Family MILLEPORIDIIDAE Yabe and Sugiyama, 1935

Genus *Milleporidium* Steinmann, 1903

Milleporidium darwini Hamilton, n. sp.

(Plate 7, figure 1)

Description.—Coenosteum calcareous, massive. One kind of distinct zooidal tube present. Coenosteum reticulate in tangential section; interspaces same general size as zooidal tubes and could be taken for tubes which are numerous and closely appressed. In longitudinal section the appearance is of parallel tubes. Tabulae few.

Dimensions.—Tubes 0.26–0.32 mm in diameter. In longitudinal section tubes are parallel with about 9 tubes and interspaces in 2 mm. Coenosteum 21 mm high and 52 mm by 15 mm wide.

Type locality.—Hess Guyot (MP 33D). Two specimens available for study; well preserved in a matrix of phosphatized limestone intercalated with manganese dioxide fragments.

Geologic age.—Association with the other fauna indicates Cretaceous (Aptian to Cenomanian) age.

Discussion.—This species is most closely related to *Milleporidium steinmanni* Yabe and Sugiyama and *M. arinokiens* Yabe and Sugiyama, but differs from the former by fewer tabulae and the latter by the growth form.

This species was named after Charles Darwin.

Milleporidium davisi Hamilton, n. sp.

(Plate 7, figure 3; Plate 9, figure 1)

Description.—Coenosteum calcareous, dendritic; apparently encrusting about a central particle of phosphatized limestone. Zooidal tubes and interspaces about the same size and curving outward from a common center, becoming nearly vertical near the surface; tubes increasing slightly near the top. Sides of the tubes minutely papillate at the top edge. Tabulae few.

Dimensions.—Zooidal tubes average 0.47 in diameter. At the surface tubes and interspaces average 5 in 2 mm. Coenosteum (including central foreign particle) 15 mm by 30 mm.

Type locality.—Hess Guyot (MP 33B). One specimen available for study; well preserved as encrusting growth around a phosphatized limestone pebble center.

Geologic age.—Association with the other fauna indicates Cretaceous (Aptian to Cenomanian) age.

Discussion.—This species seems most closely related to *Milleporidium arinokiens* Yabe and Sugiyama of the Jurassic of Japan, but has much larger zooidal tubes and fewer tabulae. This species differs from *M. darwini* by the growth form and larger tubes.

Remarks.—Yabe and Sugiyama (1935, p. 160) give the geologic range of this genus as Triassic to Cretaceous (Danian).

This species was named after W. M. Davis.

Phylum MOLLUSCA

Class GASTROPODA

Order CTENOBRANCHIATA Schweigger

Superfamily TAENIOGLOSSA Bouvier

Family VERMICULARIIDAE Zittel

Vermicularia LAMARCK, 1799. Mem. Soc. Hist. Nat., Paris, 78 *Vermetus* ADANSON, 1757 (in part of authors)

Type species (by monotypy): *Serpula lumbricalis* Linnaeus

Generic description.—Shell unattached, first whorls turretelloid, later ones irregularly coiled, surface bearing spiral ridges and also growth lines; aperture generally round.

Vermicularia ? sp.

(Plate 8, figure 2; Plate 9, figure 4)

Description.—Fragments only; 2½ whorls maximum number preserved. Initial coils tightly curled later becoming irregularly uncoiled with whorls not in contact; later whorls regularly uncoiled; shell wall thick. Whorls strongly keeled above with keel at top of whorl; in later whorls keel moves to central position and becomes less pronounced. Aperture not preserved but probably round. The empty shells may have been occupied by other organisms; they now appear as coiled channels rather than closed tubes.

Dimensions.—On fragments with best-preserved initial whorls the height of whorl initially is about 8–10 mm; later whorls on largest fragment about 25 mm high. Width expanding as whorls enlarge; width of largest fragment about 27 mm at bottom.

Type localities.—MP 33C, MP 33D, MP 33I. Large numbers of fragments available for study. Fragments cemented together by fine-grained calcareous ooze. Manganese dioxide coatings cover both shell fragments and calcareous ooze. These fossils were dredged near the top, break in slope, and below break in slope on Hess Guyot. They must have existed in vast numbers at one time.

Geologic age.—At MP 33C (Hess Guyot) these gastropods were dredged with large, rounded fragments of indurated globigerina ooze which contains an excellent foraminiferal fauna of Paleocene age. (*See* section on Foraminifera.) Within shells of the gastropods the same fauna was well preserved. By association, therefore, these fossil gastropods are probably Late Cretaceous-Paleocene with the latter more probable.

Discussion.—Fragments available do not justify description of a new species.

Remarks.—The range of the genus *Vermicularia* is reported as Upper Cretaceous to Recent (Shimer and Shrock, 1944, p. 491).

Many authors have assigned similar species to *Vermetus* Adanson 1757. Under the International

Code this name is not available in that it is prior to Linnes Systemae Naturae, 10th ed. (Jan. 1, 1758) which is taken as the starting point for the Law of Priority.

In 1799 Lamarck named the genus *Vermicularia* into which these fossils appear to fall. Daudin's *Vermetus* 1800 has been confused by many authors with *Vermicularia*. Dr. A. Myra Keen, who has made a special study of Vermetids, thinks that *Vermetus* Daudin, 1800, is a valid genus and agrees that the above fossils are probably *Vermicularia*. Dr. Keen noticed that these fossils were not closed tubes, closely and loosely wound, but were merely channels, or half tubes, wound as to simulate the usual vermiculid coiling. Dr. Keen thinks that possibly the empty shells of *Vermicularia* were occupied by a worm or other animal which resorbed or damaged the original shell. Dr. Keen notes similarities also with *Laxispira* Gabb, 1876, but thinks that both *Laxispira* and *Vermicularia* should be allocated to the Turritellidae.

Gastropod Fragments

Family NERINEIDAE Zittel

Genus *Nerinea* Defrance, 1824

Nerinea DEFRANCE, 1824. Bull. Univ. Sci. Nat. 284.

Generic description.—Turreted or pyramidal, usually imperforate. Columella invariably, and inner and outer lips generally, with simple folds.

Nerinea sp.

Description.—Two whorls in poor preservation. Whorls concave, probably smooth, with sharp keel. Columella poorly preserved but probably folded.

Locality and remarks.—Locality: Cape Johnson Guyot (MP 37A). This specimen was a constituent part of a partially phosphatized manganese dioxide coated fragment which also contained *Montastrea menardi*.

Nerinea was a common associate of the reef-coral-rudistid faunas of the Cretaceous. The genus ranges from the Jurassic to the Upper Cretaceous (Senonian).

Family CERITHIIDAE Menke

Genus *Cerithium* Brugiere, 1789

Cerithium BRUGIERE, 1789. Ency. Meth. (Vers) Vol 1 (2), XV, p. 546.

Type species: *Cerithium nodulosum* Brugiere.

Generic description: Turreted, imperforate; aperture oblong, ovate with backwardly curved canal. Outer lip often somewhat reflected. Columella concave, frequently with one or two folds.

Cerithium sp.

(Plate 6, figure 3)

Description.—Shell turreted; whorls concave with a row of small nodes at center; keels nodose; aperture not well preserved but was oblong. Canal not preserved.

Dimensions.—Width at base 0.83 mm; height about 1.8–2.0 mm.

Locality and remarks.—Locality: Cape Johnson Guyot (MP 37C). Fossil stained orange and cemented to surface of detritus and could have been later in age than constituent fossils of the inside portion. The range of the genus is Jurassic to Recent.

Family TROCHIDAE Adams

Genus Trochus Linnaeus, 1758

Trochus LINNAEUS, 1758. Syst. Nat. 10th ed., p. 756.

Type species: *Trochus maculatas* Linnaeus.

Generic description: Shell conical or pyramidal. Whorls slightly convex or flat. Base angular at the periphery. Inner lip often truncated anteriorily, thickened, or with teeth.

Trochus sp.

(Plate 5, figure 4)

Description.—Small shell, trochoid with depressed sutures and flat whorls. Sculpture on whorls: symmetrical rows of large nodes. Base of shell and aperture not preserved.

Locality and remarks.—Locality: Cape Johnson Guyot (MP 37C). Fragment preserved in fossiliferous, manganese-dioxide-coated debris. The genus ranges from Silurian to Recent and was common in the reef coral-rudistid faunas of the Cretaceous.

Phylum MOLLUSCA

Class PELECYPODA

Order TELEODESMACEA Dall

Superfamily CHAMACEA Geinitz

Family CAPRINIDAE d'Orbigny

Genus *Caprina* d'Orbigny

Caprina mulleri Hamilton, n. sp.

(Plate 8, figure 3)

Description.—Fixed valve unknown. Free valve spirally twisted ½ turn; shell thick and middle layer perforated by numerous simple wide parallel canals radiating from the umbo. Canal walls, or radial plates, nonbifurcating with the exception of three or four radial plates in the anterior margin. Posterior dentition not preserved, anterior tooth large, supported by a plate which divides the umbonal cavity lengthwise. Hinge margin thickened and grooved for cartilage. Surface of shell marked by lines radiating from the umbo owing to weathering of outer layer and exposure of the radial plates.

Dimensions.—Length 90 mm. Width greatest near ventral margin: about 68 mm. Thickness (convexity) about 45 mm. Canals about 6 mm high and 1.5–2.0 mm wide.

Type locality.—Cape Johnson Guyot (MP 37C). One well-preserved, phosphatized specimen available for study. Specimen was a detrital shell in a matrix of partially phosphatized shell and calcareous material with many fragments and coatings of manganese dioxide. This specimen, dredged from below break in slope on Cape Johnson Guyot; probably rolled down slope from living position.

Geologic age.—The range of the genus *Caprina* is from Aptian to Turonian. The range of the family Caprinidae is from Aptian to Turonian (Palmer, 1928, p. 58). Paquier (1903, p. 71) believed that the nonbifurcating canal walls are more primitive than the bifurcating type. Palmer (1928, p. 21) noted that no bifurcation of radial plates is reported below the Cenomanian, and subsequent to that period simple plates are rare. The genus occurs sparsely below the Cenomanian and during that period had a rapid expansion which ended at the end of the Cenomanian. All of the caprinid fragments are from the genus *Caprina* and belong to the earlier evolutionary form in which the canal walls are nonbifurcating though a few walls of the larger valve show some bifurcation. The probable age of this species and of the other valves and fragments can definitely be confined to the range Aptian to Turonian, and probably the mollusks lived in the range Aptian to Cenomanian. No further restriction of the range is advisable, though the affinities of *C. mediopacifica*, described next are with the Aptian primitive caprinid genus *Praecaprina*.

Discussion.—This species has no close affinities with described species but is clearly a representative of the genus *Caprina*.

Remarks.—This species was named after Dr. Siemon W. Muller of Stanford University. The distribution of the genus *Caprina* was in the warm, shallow waters of the Tethys Sea and is known from Southern England, Europe, the Mediterranean area, Texas, and Mexico. In these areas the rudists are closely associated with coral reefs as is the case in the Mid-Pacific area of this paper.

Caprina mediopacifica Hamilton, n. sp.

(Plate 8, figure 1)

Description.—Fixed valve unknown. Free valve spirally twisted about ⅓ turn; shell thick and middle layer perforated by numerous, simple, tear-drop-shaped parallel canals radiating from the umbo. Canal walls, or radial plates, nonbifurcating. In the center of ventral margin the canals become small, oblique, and obscure. Posterior dentition obscure. Anterior tooth blunt, supported by a plate which divides the umbonal cavity lengthwise. Anterior margin with distinct ligamental groove. Surface of shell marked by lines radiating from the umbo owing to exposure of radial plates through weathering. Shell distinctly wider ventrally.

Dimensions.—Length about 34 mm. Width (of ventral portion) about 27 mm. Thickness (convexity) about 17 mm. Canals drop-shaped and largest ones about 3 mm high.

Type locality.—Cape Johnson Guyot (MP 37C).

Geologic age, discussion and remarks. (In addition to items under *C. mulleri*)—This species is allied to the Aptian genus *Praecaprina* Paquier which is the primitive caprinid and believed by Paquier (1903 p. 81, 82) to be the ancestor of *Caprina*. Paquier noted (1903, p. 72), "La région ventrale des *Praecaprina* reste toujours dépourvue de canaux; mais on a vu que chez la première des Caprines, leur présence dans cette région apparait comme une acquisition récente."

The canals of the central ventral margin of this species are small, oblique, and obscure but nevertheless are present; thus this species must be placed in the genus *Caprina*. It is probably a very early primitive caprinid which, together with the nonbifurcating radial plates of the remainder of the caprinid material from the Cape Johnson Guyot, might argue for an occurrence early in the range Aptian-Cenomanian as discussed under *C. mulleri*.

Caprina fragments

Discussion.—There are many shell fragments of members of the family Caprinidae and the genus *Caprina* in the fossil material from Cape Johnson Guyot (MP 37C). This material is too poorly preserved for description, but indicates the abundance of rudists in the reef coral-rudistid fauna. One fragment of a shell shows very large canals which are parallel, about 15–18 mm high, and separated by thin, nonbifurcating walls. This fragment resembles closely the canal structures of *Caprina chofati* described by Douville (1898, p. 143–147) from the Upper Albian of Portugal.

Superfamily CARDITACEA Menke

Family CARDITIDAE Gill

Genus *Cardita* Bruguiere

Cardita Bruguiere, 1792. Ency. Meth. (Vers) (2), 401.

Type species: *Cardita* ss. Lamarck 1799; type *C. calyculata* Linne.

Generic description: Elongate, quadrate, with prominent very anterior beaks; sculpture radiate and usually imbricate; commonly with a lunule; inner margin dentate; cardinals long and oblique. Trias to Recent.

Cardita sp.

Description.—Shell fragments only. Shell quadrate; beak anterior; sculpture radiate and imbricate; cardinals long and oblique.

Locality and remarks.—Locality: Cape Johnson Guyot (MP 37C). The shell was cemented to the outside of a detrital limestone mass and may have lived at a later time than the constituent parts of the enclosed fossils.

Phylum ECHINODERMATA

Class ECHINOIDEA

Order EXOCYCLOIDA

Family ECHINONEIDAE Wright, 1864

Genus *Pyrina* Desmoulins, 1835

Pyrina DESMOULINS, 1935. Etude sur Les Echinides, Bordeaux (1835–1837).

Type species (genolectotypes in dispute): Lambert, 1908:

Pyrina petrocoriensis Desmoulins. Cooke, 1946: *Nucleolites castana* Brongniart, 1822. (*See* Remarks).

Generic description: Form generally regular, circular, or oval, more or less swollen on top; inferior surface slightly concave; summit central; four genital pores; interambulacral areas double or triple ambulacral areas; ambulacras complete, straight, flat, border of each single pairs of close pores, united or not by a ridge; tubercles of two kinds: milaires more numerous above than below; papillaires more numerous but a little larger below than above; spines unknown; mouth symmetrical, central, round, sunken oblique; anus supra-marginal, or between the border and the summit, oval or round, occasionally large.

Pyrina keenae Hamilton, n. sp.

(Plate 9, figure 6)

Description.—Test oval; upper surface inflated; under surface inflated, depressed around the mouth; oral side curved and concave from side view. Highest point of test at apical system and central. Apical system not preserved. Mouth central, symmetrical, oblique, oval. Anus supramarginal, pyriform, just above ambitus. Ambulacral areas outlined by pairs of pores, raised aborally, depressed orally and extending from the peristome to apical system; widest at ambitus. Pores in pairs, nonconjugate and separated by costae. Whole surface covered by small granules and mammilated tubercles. Tubercles are imperforate and larger on oral side.

Dimensions.—Length: 33 mm. Width: 26 mm. Height: 20 mm.

Type locality.—MP 33K. One specimen available for study; poorly preserved aborally, but well preserved orally under a thin coating of manganese dioxide. Position of the specimen was below the break in slope among reef coral debris.

Geologic age.—Association with the other fossils indicates Cretaceous (Aptian to Cenomanian) age.

Discussion.—This species appears to be most closely related to *P. incisa* d'Orbigny, 1856 (spelled *P. insisa* in Plate fig.) but differs in size, position of anus, size of tubercles, and in lacking depressed area around anus shown in figure of *P. incisa*.

Remarks.—Desmoulins (1835) originally named the genus *Pyrina* and assigned to it seven fossil species: *Pyrina echinonea, Echinoneus cassidularis, Nucleolites castanea* Brongniart, *Nucleolites depressa* Brongniart, *Galerites rotula* Brongniart and two "new species" which he did not name. The first two species named were doubtfully assigned. According to the International Code of Zoological Nomenclature the two "new species" of the original description are *nomina nuda*. (As pointed out by Cooke, 1946, p. 220.)

In 1908 Lambert designated *P. petrocoriensis* as the type species of *Pyrina*. Cooke (1946, p. 220) discarded Lambert's designation since it was a *nomen nudum* when the genus was first described and designated *Nucleolites castanea* Brongniart, 1822, as the type species, because it was the most abundant of the three species definitely assigned to the new genus (*Pyrina*) by Desmoulins. Cooke calls attention to the fact that Cotteau (d'Orbigny, 1853–1860, p. 474, footnote) identified *Nucleolites castanea* as the species figured under the name *Echinoconus castanea* by d'Orbigny (1853–1860, p. 503, Pl 990).

Cotteau examined the original material of Brongniart and in the footnote cited stated (author's translation): "*Nucleolites castanea* appears to me, as to the majority of other authors, as only a round and oblong variety of *Echinoconus castanea* characteristic of the Albian stage." The figure of *Echinoconus castanea* (d'Orbigny, 1853–1860, Pl 990) clearly shows the anus on the oral side. Desmoulins clearly stated in his description of the new genus *Pyrina* that the anus is supramarginal or on the aboral side.

The type species selected by Cooke (*Nucleolites castanea* Brongniart, 1822) is not a true *Pyrina* as understood by most authors for a hundred years and is better excluded from that genus rather than selected as a type species. Although Desmoulins did not name his "two new species" it is clear from his later publications that these were *P. petrocoriensis* and *P. dubia*, and that they were probably used to help describe the new genus. Lambert's designation of *P. petrocoriensis*, while perhaps not strictly in accordance with the rules of the International Code, is probably a true representation of Desmoulin's original intentions. On the other hand there is no doubt that Cooke (1946) acted within the International Code. Mortensen (1948) discussed these problems at some length and concludes that *Globator* Agassiz is a synonym of *Pyrina* and that *P. petrocoriensis* is the type species of the genus *Pyrina*. This problem should be given further study and perhaps presented to the International Commission on Zoological Nomenclature as a case meriting suspension of the Rules.

The range of the genus *Pyrina* is Late Jurassic to Eocene. The genus reached its culmination in France in the Senonian and practically died out everywhere by the end of the Cretaceous. The genus is known from England, Europe, Africa, India, Mexico, and Brazil. Neaverson (1928, p. 413) notes the highest occurrence of *Pyrina* in England as Cenomanian and believes that the genus was a shallow-water form confined to littoral tracts.

This species was named after Dr. A. Myra Keen of Stanford University.

PART 2: FAUNAL LISTS OF FORAMINIFERA

Cores at MP 27.—The mixed planktonic fauna of the cores from MP 27 are divided into two parts for clarity: the Upper Cretaceous fauna and the Tertiary fauna. The modern tropical Pacific planktonic species are not listed because they are well described and illustrated in the current literature. Further discussion and illustrations are reported by Hamilton (1953).

Upper Cretaceous fauna:

Globigerina cretacea d'Orbigny
Globigerinella aspera (Ehrenberg)
Globorotalia velascoensis (Cushman)
Globotruncana arca (Cushman)
G. calcarata Cushman
G. caliciformis (de Lapparent)
G. canaliculata (Reuss)
G. contusa (Cushman)
G. fornicata Plummer
G. gansseri Bolli
G. globigerinoides Brotzen
G. marginata (Reuss)
G. rosetta (Carsey)
G. stuarti (de Lapparent)
G. ventricosa (White)
Gümbelina costulata Cushman
G. excolata Cushman
G. globulosa (Ehrenberg)
G. striata (Ehrenberg)
G. plummerae Loetterle
G. pseudotessera Cushman
G. ultimatumida White
Pseudotextularia varians Rzehak
Rugoglobigerina rugosa (Plummer)
Ventilabrella austinana Cushman
V. carseyae Plummer

Tertiary to Recent fauna:

In addition to the modern tropical Pacific species the following Tertiary planktonic species have been identified:
Globigerina altispira Cushman and Jarvis
G. bulloides d'Orbigny
G. cretacea d'Orbigny
G. pseudobulloides Plummer
G. triloculinoides Plummer
G. venezuelana Hedberg
Globigerinella voluta (White)
Globigerinoides mexicana (Cushman)

Globorotalia aragonensis Nuttall
G. crassata (Cushman)
G. crassata (Cushman) var. *aequa* Cushman and Renz
G. crassata densa (Cushman)
G. velascoensis (Cushman)
G. velascoensis (Cushman) var. *acuta* (Toulmin)
G. wilcoxensis Cushman and Ponton
Hantkenina alabamensis Cushman
Sphaeroidinella multiloba LeRoy
S. rutschi Cushman and Renz
S. seminulina (Schwager)
Eponides trümpyi Nuttall (benthonic)

Fauna on Hess Guyot.—The following planktonic species have been identified from the indurated globigerina ooze from MP 33 C:

Globigerina aff. *G. angulata* White
G. bulloides d'Orbigny
G. compressa Plummer
G. cretacea d'Orbigny
G. eocaena Gümbel
G. pseudobulloides Plummer
G. triloculinoides Plummer
Globigerinella pseudovoluta Bandy
Globorotalia crassata Cushman
G. aff. *G. inconspicua* Howe
G. velascoensis (Cushman)
G. velascoensis (Cushman) var. *acuta* Toulmin)
Gümbelina aff. *G. glabrans* Cushman
G. globulosa (Ehrenberg)
G. ultimatumida White
Eponides trümpyi Nuttall

Fauna on Horizon Guyot.—The following planktonic species have been identified from the Lower Eocene fauna from the two cores (MP 25 E-1 and E-2) from Horizon Guyot; the same fauna was present in both cores with little variation:

Globigerina apertura Cushman
G. bulloides d'Orbigny
G. aff. *G. eocaena* Gümbel
G. linaperta Finlay
G. mckannai White
G. ouachitaensis Howe and Wallace
G. ouachitaensis Howe and Wallace var. *senilis* Bandy
G. orbiformis Cole
G. rotundata d'Orbigny var. *jacksonensis* Bandy
G. triloculinoides Plummer
Globigerinoides nuttalli Hamilton, n. sp.
G. mexicana (Cushman)
Globigerinella pseudovoluta Bandy
Globorotalia aragonensis Nuttall
G. centralis Cushman and Bermudez
G. crassata (Cushman)
G. crassata densa (Cushman)
G. aff. *G. inconspicua* Howe
Gümbelina cubensis Palmer
G. venezuelana Nuttall
Hantkenina mexicana Cushman var. *aragonensis* Nuttall
Hastigerinella eocanica Nuttall
H. eocanica aragonensis Nuttall
Eponides trümpyi Nuttall (benthonic)

Fauna on Guyot 19171.—The following planktonic species have been identified from the indurated globigerina-ooze boulder taken by dredge (MP 26 A-3) across the break in slope on Guyot 19171:

Globigerina triloculinoides Plummer
Globorotalia aragonensis Nuttall
G. centralis Cushman and Bermudez
G. crassata (Cushman)
G. crassata densa (Cushman)
G. wilcoxensis Cushman and Ponton

FIGURE 1. DREDGE HAUL FROM MP 25 F-2
Manganese-coated volcanic rocks and manganese nodules.

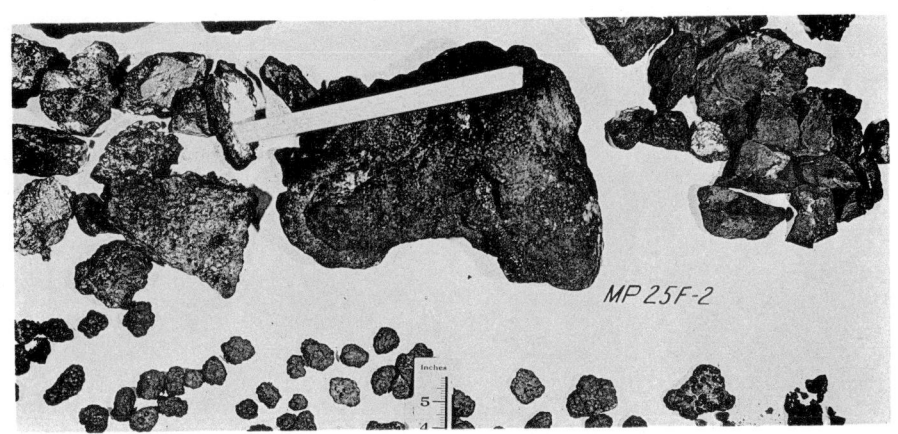

FIGURE 2. DREDGE HAUL FROM MP 26 A-3 Top right and left: olivine basalt boulders; top center: manganese-coated, indurated Globigerina ooze; bottom: manganese nodules and slabs of olivine basalt.

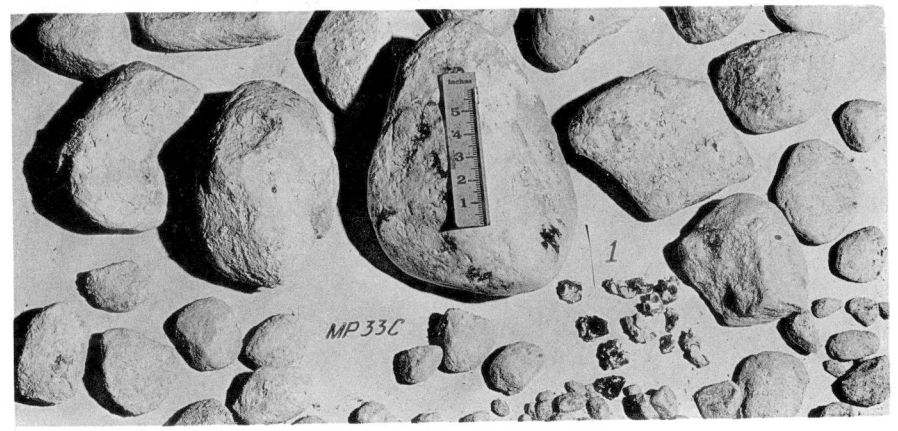

FIGURE 3. DREDGE HAUL FROM MP 33 C
The rounded rocks are soft but indurated Globigerina ooze. Rounding probably took place within the dredge.

DREDGE HAULS FROM THE FLAT-TOPPED SEAMOUNTS

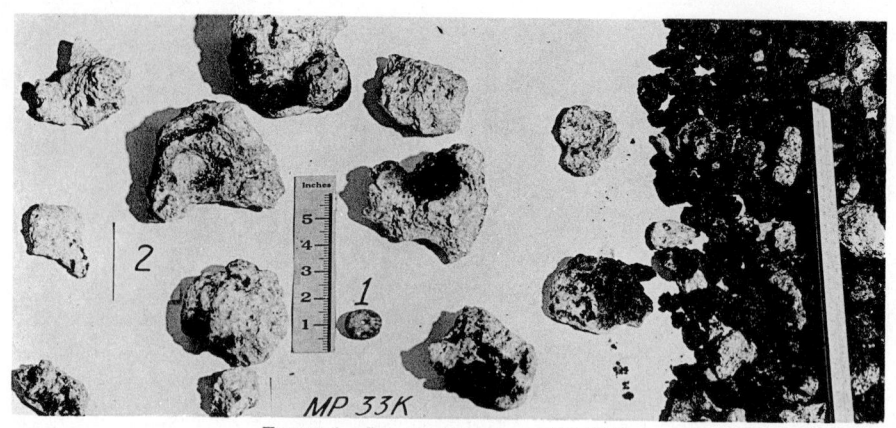

FIGURE 1. DREDGE HAUL FROM MP 33 K
(1) Is an irregular echinoid. Rocks, including (2), are reef coral.

FIGURE 2. DREDGE HAUL FROM MP 37-A
The large rock (top center) is sandstone covered by manganese. Some of the smaller pieces are reef coral

FIGURE 3. DREDGE HAUL FROM MP 37-C
Fossiliferous limestone and calcareous ooze which has been partially phosphatized and coated with manganese; *note* the well-preserved rudistid to the left of the 6-inch rule.

DREDGE HAULS FROM THE FLAT-TOPPED SEAMOUNTS

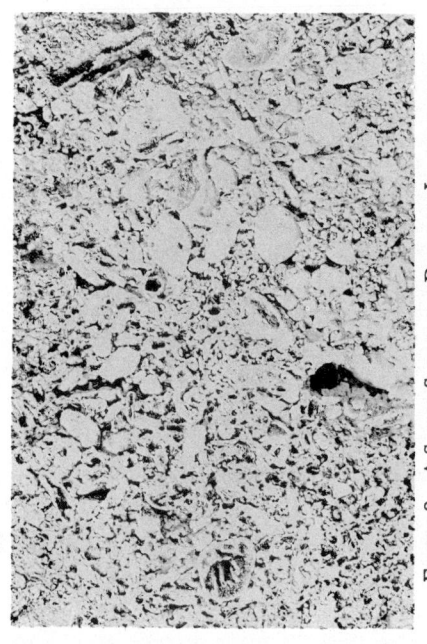

FIGURE 1. BASALTIC GRAVEL AND SAND MIXED WITH RED CLAY
Sample from the bottom of core MP 27-2 from a depth of 2050 fathoms; × 1

FIGURE 2. A SAWED SECTION OF DETRITAL LIMESTONE FROM MP 33-A; × 7

FIGURE 3. ROCK TYPES FROM MP 33-C
(a) Coquina of gastropod fragments cemented with calcareous ooze, × 2; (b) Rounded olivine basalt pebble, × 2½; (c) Section of manganese nodule with a phosphatized limestone center, × 2

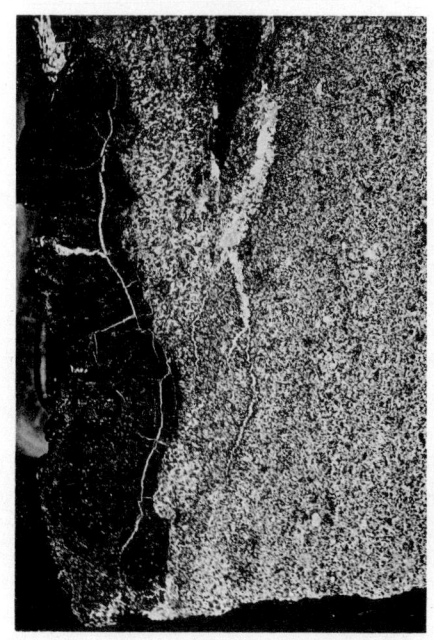

FIGURE 4. SANDSTONE FROM CAPE JOHNSON GUYOT (MP 37-A)
Note the coat of manganese at the top; × 4½

ROCK TYPES FROM THE DREDGE HAULS

GEOL. SOC. AM., MEMOIR 64 HAMILTON, PL. 5

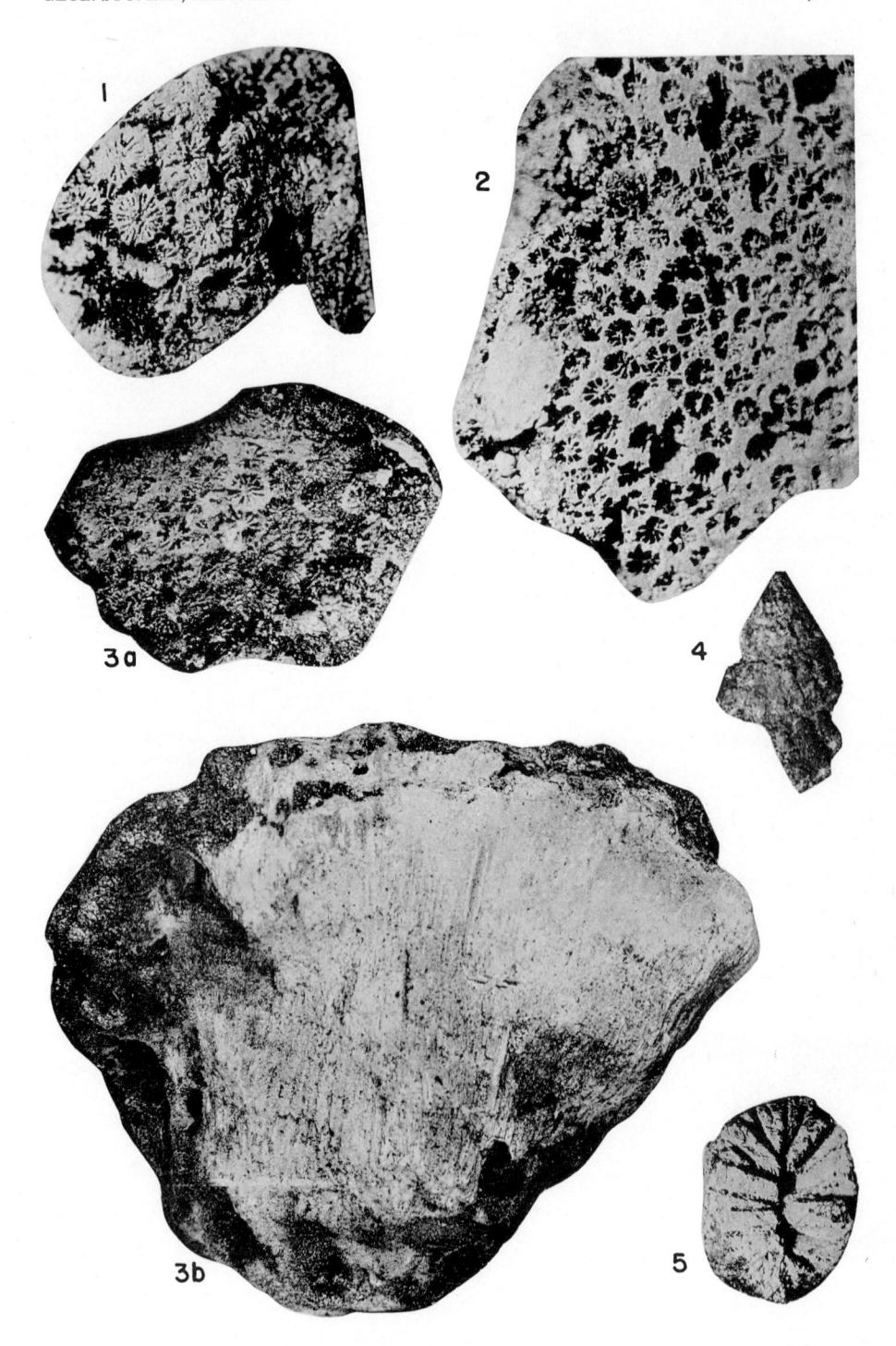

FOSSILS FROM DREDGE HAULS

APPENDIX A. SYSTEMATIC PALEONTOLOGY

PLATE 5.—FOSSILS FROM DREDGE HAULS

Figures / Page

1. *Montastrea menardi* Hamilton, n. sp... 59
   Holotype; top view; Hess Guyot (MP 33 K). × 3.
2. *Astrocoenia dietzi* Hamilton, n. sp... 57
   Holotype; top view; Cape Johnson Guyot (MP 37 A). × 3.3.
3. *Astrocoenia revellei* Hamilton, n. sp... 57
   Holotype; a, top view, × 3; b, side view, × 1; Hess Guyot (MP 33 K).
4. *Trochus* sp... 65
   Side view of fragment; Cape Johnson Guyot (MP 37 C). × 3.4.
5. *Lophosmilia fundimaritima* Hamilton, n. sp... 60
   Holotype; top view; Cape Johnson Guyot (MP 37 C). × 2.

PLATE 6.—FOSSILS FROM DREDGE HAULS

Figures | Page
1. *Brachyseris wellsi* Hamilton, n. sp.. 58
    Holotype; a, top view; b, side view; Hess Guyot (MP 33 K). × 3.
2. *Brachyseris montemarina* Hamilton, n. sp... 58
    Holotype; a, top view; b, side view; Cape Johnson Guyot (MP 37 C). × 3.
3. *Cerithium* sp.. 64
    Side view of fragment; Cape Johnson Guyot (MP 37 C). × 20.
4. *Microsolena* sp... 59
    Top view; Hess Guyot (MP 33 K). × 3.
5. *Diploastrea* sp... 60
    Top view; Hess Guyot (MP 33 K). × 3.

GEOL. SOC. AM., MEMOIR 64　　　　　　　　　　　　　　　　HAMILTON, PL. 6

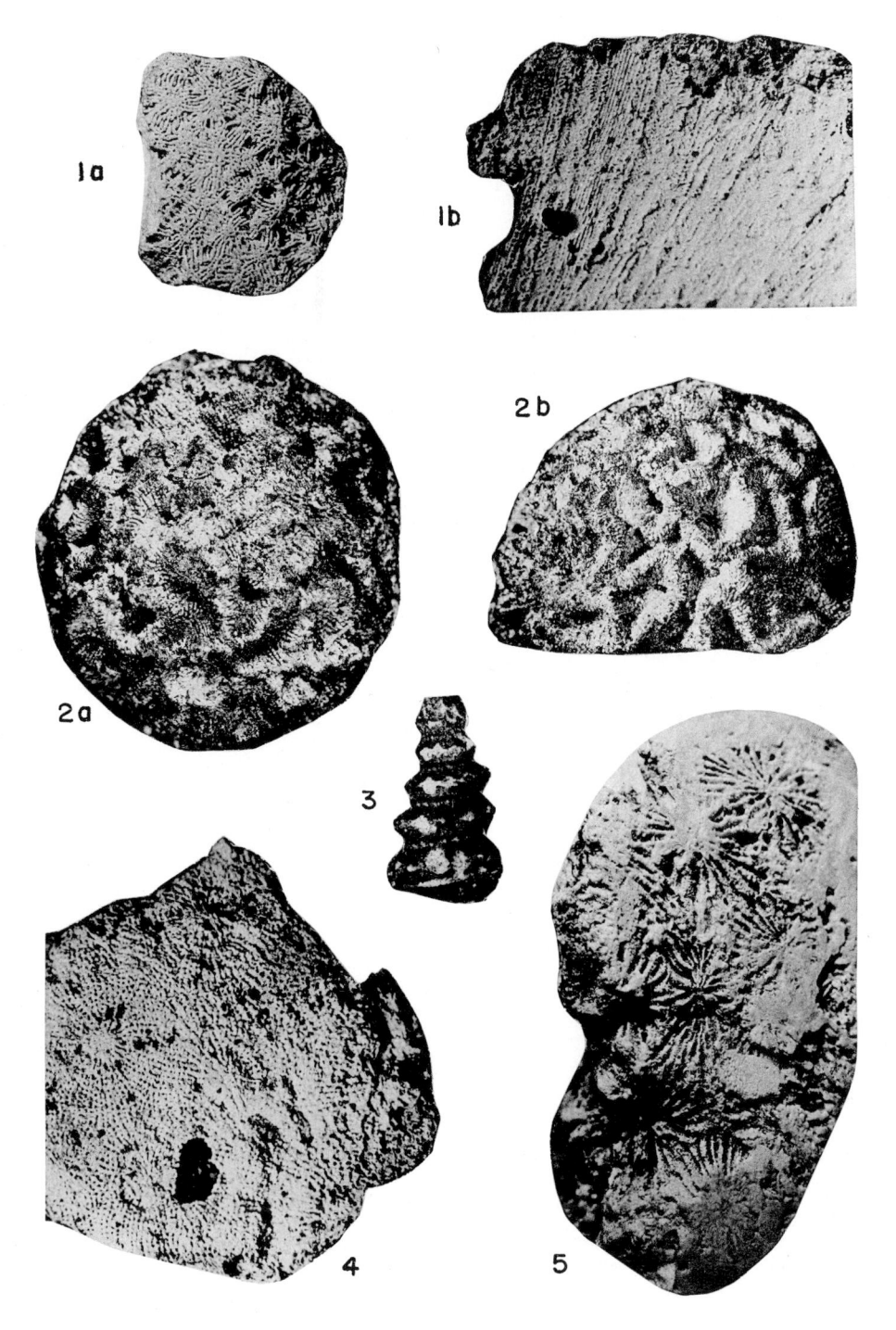

FOSSILS FROM DREDGE HAULS

FOSSILS FROM DREDGE HAULS

APPENDIX A. SYSTEMATIC PALEONTOLOGY

PLATE 7.—FOSSILS FROM DREDGE HAULS

Figures	Page
1. *Milleporidium darwini* Hamilton, n. sp.	62
Holotype; a, side view; b, top view; Hess Guyot (MP 33 D). × 4.	
2. *Stromatopora* sp.	62
Side view; Hess Guyot (MP 33 K). × 4.3.	
3. *Milleporidium davisi* Hamilton, n. sp.	62
Holotype; top view; Hess Guyot (MP 33 B). × 4.3.	
4. *Actinostroma pacifica* Hamilton, n. sp.	61
4a: holotype; side view; Cape Johnson Guyot (MP 37 C). × 3.4.	
4b: side view; Hess Guyot (MP 33 K). × 3.	

PLATE 8.—FOSSILS FROM DREDGE HAULS

Figures / Page

1. *Caprina mediopacifica* Hamilton, n. sp.. 66
   Holotype; a, side view; b, interior view; Cape Johnson Guyot (MP 37 C). × 2.
2. *Vermicularia* sp.. 63
   Side view of lower whorls; Hess Guyot (MP 33 D). × 1.7.
3. *Caprina mulleri* Hamilton, n. sp.. 65
   Holotype; view of interior; Cape Johnson Guyot (MP 37 C). × 1.6.

GEOL. SOC. AM., MEMOIR 64　　　　　　　　　　　　　　　HAMILTON, PL. 8

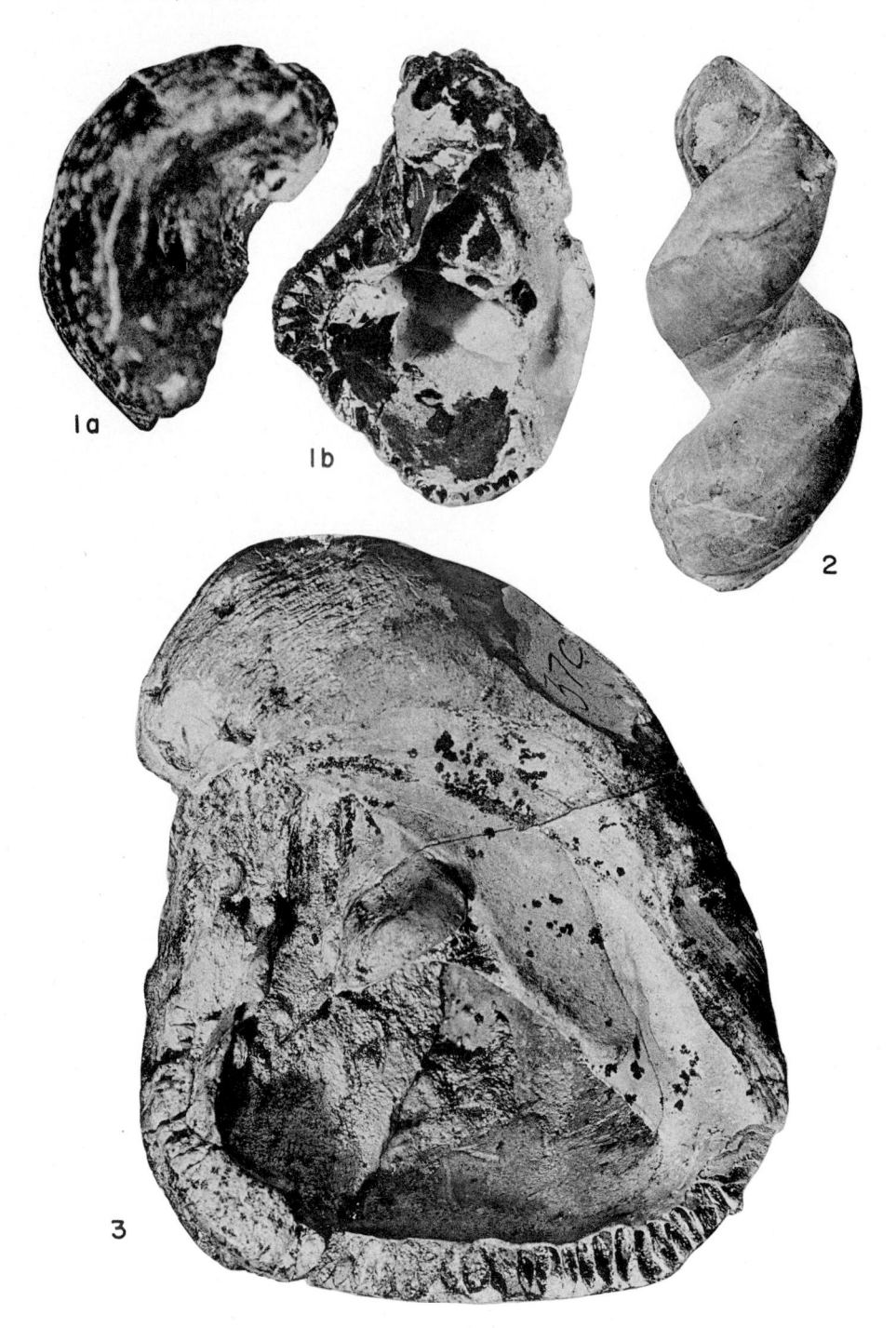

FOSSILS FROM DREDGE HAULS

GEOL. SOC. AM., MEMOIR 64　　　　　　　　　　　　　　　　HAMILTON, PL. 9

FOSSILS FROM DREDGE HAULS

APPENDIX B. LITHOLOGY AND SEDIMENTS

The following descriptions are supplementary to the general descriptions included in the first part of this paper.

MP 25F-1.—(For description of locality see "Geology of the Guyots," "Horizon Guyot".) Total haul composed of three large (4 by 8 by 2 inches) deeply weathered basaltic rocks completely coated over with manganese dioxide up to 35 mm in thickness with fossil and Recent Foraminifera in interstices, and many small chips of manganese-coated basaltic rock. Small manganese pellets are interspersed through the calcareous material. One manganese nodule was sectioned and study of the thin section revealed that the center was of phosphatized limestone.

MP 25F-2.—Recognized in this dredge haul were:

(1) Fresh and altered chips of olivine basalt covered with manganese dioxide coatings up to 30 mm in thickness.

(2) Fossil globigerina ooze coated with manganese dioxide.

(3) Large manganese-coated boulder of broken basalt fragments cemented together by globigerina ooze, calcium carbonate, and light-brown clay coated with manganese dioxide up to 10 mm in thickness and containing many manganese nodules.

(4) Many manganese nodules. Thin section and hand specimen studies revealed the following centers:

(a) Subrounded pebble of phosphatized limestone, 7 by 10 mm.

(b) Subrounded olivine basalt pebble, 10 by 14 mm; minerals: olivine completely altered to iddingsite, augite, labradorite; texture: intersertal.

(c) Chalcedony replacing calcite center; 11 by 10 mm, subangular.

(d) Olivine basalt pebble showing definite signs of wear; subrounded; 20 by 15 mm; minerals: olivine completely altered to iddingsite, labradorite, opaques (magnetite and ilmenite), apatite, augite, calcium carbonate in holes; texture intergranular, holocrystalline, amygdaloidal.

(e) Subrounded olivine basalt pebble, 6 by 8 mm; minerals: labradorite, iddingsite altered from olivine, augite, opaques; texture: intergranular, amygdaloidal.

(f) Subrounded olivine basalt pebble, 24 by 15 mm; minerals: olivine with sagenitic webs, labradorite, serpophite (in altered olivine), opaques, chrysotile alteration, large crystals of augite; texture: ophitic.

(g) Subrounded pebble of indurated, partially phosphatized calcareous material; minerals; labradorite, apatite, organic material (Foraminifera or Radiolaria) now phosphatized.

(h) Olivine basalt chip; minerals: labradorite, olivine altered to iddingsite, augite; texture: intergranular.

MP 26A-1.—A few grams of manganese dioxide chips and foraminiferal sand; the core barrel of the Phleger Bottom Sampler was badly smashed indicating that the corer probably hit a rock bottom.

MP 26A-3.—Observed in this dredge haul were:

PLATE 9.—FOSSILS FROM DREDGE HAULS

Figures — Page

1. *Milleporidium davisi* Hamilton, n. sp. 62
    Holotype; side view of section; organism encrusted around rock center; Hess Guyot (MP 33 B). × 3.
2. Ahermatypic coral of the Subfamily Oculininae Vaughan and Wells; top view; MP 34. × 4. — 61
3. Earbone of cetacean. Cape Johnson Guyot (MP 37 B). × 2. 29
4. *Vermicularia* sp. 63
    Side view of upper whorls; Hess Guyot (MP 33 D). × 1.7.
5. Shark's tooth (*Oxyrhina?* sp.) 28
    Side view; tooth is center of manganese nodule; Guyot 19171 (MP 26 A-3). × 2.
6. *Pyrina keenae* Hamilton, n. sp. 67
    Holotype; a, side view; b, oral view; c, aboral view; Hess Guyot (MP 33 K). × 2.4.

(1) Basalt boulder: 13 by 8.5 by 6 inches, covered with 6-10 mm of manganese dioxide; corners and edges definitely rounded; a thin section of a chip from this boulder was of olivine basalt.

(2) Basalt boulder: 8.5 by 6 inches; manganese coatings up to 8 mm; corners and edges rounded; chip revealed the boulder to be of amygdaloidal olivine basalt with relatively large phenocrysts of feldspar.

(3) Round, flat (14½ inches across and 6 inches thick) boulder of indurated globigerina ooze covered with 6 mm of manganese dioxide on top and discoloration of manganese only on the bottom.

(4) A large number of manganese nodules; about 25 of these were sawed through or broken and the majority had centers of globigerina ooze or just a sprinkling of Foraminifera. The following nodule centers were studied in hand specimens and by thin sections:

(a) A phosphatized calcareous ooze as revealed by the outlines of altered Foraminifera.

(b) A soft, brownish-white, subrounded pebble of calcareous ooze (16 by 12 mm) which has been partially silicified. The original material was globigerina ooze as indicated by the numerous tests some of which still have a small rim of calcite left.

(c) A subangular, amygdaloidal olivine basalt pebble (15 by 10 mm); minerals: augite (large phenocrysts outlined but small remnants preserved), olivine altered to iddingsite, labradorite, calcite filling of amygdules; texture intersertal and highly altered.

(d) A composite pebble of sandstone and limestone (20 by 15 mm); minerals: detrital augite abundant, rounded opaques, rounded basaltic rock fragments, perovskite, biotite, hornblende, manganese pellets, calcium carbonate, chalcedonic silica.

(e) Soft globigerina ooze, 12 by 14 mm.

(f) Chip of olivine basalt; minerals: large augite phenocrysts, in a felted mass of laths of plagioclase, olivine minor, numerous opaques; texture intergranular.

(g) Subangular pebble of partially phosphatized globigerina ooze; outlines of many Foraminifera visible.

In addition to the above material at station MP 26 A-3 there were two slabs of olivine basalt with rounded corners which were coated with manganese crusts 4 mm in thickness. Dimensions: triangular cross section slab: 4 by 3 inches; flat slab: 3 by 1 by 2½ inches.

MP 26B.—Four or five manganese nodules and fragments of manganese were dredged at this station. A thin section of the center of one manganese nodule was an aphanitic, microcrystalline basalt.

MP 27.—Thin sections were made of ten pebbles from the bottom end of core MP 27-2. Study of these thin sections revealed the following centers:

(1) Subrounded pebble of basalt, 36 by 18 by 7 mm; minerals: labradorite, augite, no olivine; texture intergranular.

(2) Subrounded pebble of basalt, 38 by 21 by 12 mm; minerals: labradorite, augite, no olivine; texture intergranular.

(3) Rounded pebble of porphyritic basalt, 12 by 13 by 13 mm; minerals: a few very large phenocrysts of labradorite, augite; texture ophitic.

(4) Angular pebble of vesicular basalt, 20 by 16 by 7 mm; minerals: labradorite, augite; some vesicles filled with calcium carbonate; texture intersertal.

(5) Subangular pebble of amygdaloidal olivine basalt, 25 by 30 by 40 mm; minerals: labradorite, large phenocrysts of olivine, augite; texture intergranular.

(6) Subrounded pebble of amygdaloidal, porphyritic basalt, 18 by 12 by 9 mm; minerals: large phenocrysts of augite and labradorite; ground mass resembles palagonite; texture intersertal.

CORE MP 27-2P.—This core was taken by the Phleger Bottom Sampler attached to the trip arm of the piston type corer which took core MP 27-2. The other cores taken at this station have been described in the section "Cores at MP 27".

Inches from top of core	Material
0–1	Cream-colored globigerina ooze with specks of manganese; one nodule of manganese.
1–2	Cream-colored globigerina ooze with 2 light-brown red clay layers 5 mm wide and 6 mm apart.
2–6	Cream-colored globigerina ooze with scattered manganese flakes.

APPENDIX B. LITHOLOGY AND SEDIMENTS

6–7   White globigerina ooze with scattered flakes of manganese.
7–12   Red clay with abundant forams and black flakes of manganese.
12–18   Red clay with small manganese nodules.
18–22   Red clay; manganese flakes and nodules.
22–24½   White globigerina ooze with flakes of manganese.
24½–26¾   Red clay with manganese flakes and nodules.
26¾–31½   Sandy red clay with many small pebbles of subrounded to angular basalt.
31½–32½   Same as above with two 5 mm layers of globigerina ooze.
32½–33¼   Same as for 26¾–31½.
33¼–34½   Same as above with an 8 mm layer of globigerina ooze.

Complete mechanical analyses were made of four samples from the bottom of Core MP 27-2. The general conclusions to be drawn from these mechanical analyses are discussed in the section "Geology of the Guyots." The cumulative curves are illustrated in Figure 9. One mechanical analysis was made of the top 4 inches of Core MP 27-2; the cumulative curve for this analysis is illustrated in Figure 10.

MP 33A.—The dredge haul at this station yielded calcareous rock composed of rounded pellets of calcium carbonate, pelecypod shells and fragments of tests of Foraminifera cemented together by calcium carbonate. Several spongy crusts of manganese up to 13 mm in thickness were also dredged. The pipe dredges yielded globigerina ooze.

MP 33B.—The small haul at this station consisted of globigerina ooze from the pipe dredges and three stromatoporoids encrusted around rock centers of phosphatized limestone.

MP 33C.—The haul at this station consisted almost entirely of rounded, white boulders of indurated globigerina ooze ranging in size from minute particles to rocks up to 12 by 9 by 5½ inches. Some of the rocks had coatings of manganese in spots which range from a mere discoloration up to 1 mm in thickness. Also in the same haul was a small amount of coquina of broken shells of *Vermicularia* cemented together by calcareous ooze. A small number of manganese nodules were dredged as well as one well rounded basalt pebble. Thin sections of the manganese nodule centers revealed the following:

(1) Subangular pebbles of partially phosphatized limestone; many tests of Foraminifera are plainly visible.

(2) Rounded pebble of olivine basalt, 20 by 17 by 28 mm; minerals: labradorite, basaltic hornblende, olivine, augite, opaques; texture intergranular with phenocrysts of labradorite, augite, and olivine.

(3) A nodule of solid manganese dioxide.

MP 33D.—Large haul of coquina of fragments of shells of gastropods (*Vermicularia*) cemented together by finely divided calcareous ooze and with coatings of manganese up to 10 mm in thickness.

MP 33K.—Many fragments of reef building hexacorals and aggregates of calcareous debris (coralline algae, stromatoporoids, echinoid spines, casts and molds of pelecypods and pelecypod shell fragments, broken tests of Foraminifera) cemented together by calcium carbonate and covered with manganese dioxide coatings up to 16 mm in thickness. A large amount of crusts of manganese dioxide and one well-preserved echinoid (*Pyrina*) were also taken at this station. The small pipe dredges yielded globigerina ooze.

MP 37A.—The following material was dredged at this station:

(1) Fragments of two genera of reef coral, pelecypod shells and gastropod shells cemented together by calcium carbonate and manganese dioxide. Some of the calcareous material had been partially phosphatized. One side of the aggregate was covered by spongy manganese dioxide up to 5 mm in thickness.

(2) A small rock of partially phosphatized limestone composed of one well-preserved solitary coral, pelecypod shell fragments and calcareous ooze.

(3) Two thick manganese dioxide crusts up to 35 mm in thickness.

(4) A manganese-coated sandstone rock (12 by 8½ by 4 inches); manganese coat up to 45 mm in thickness; layered appearance owing to clay partings and some linear layers of manganese pellets. A thin section of this sandstone revealed the following:

Minerals: Augite, labradorite, hypersthene, hornblende, biotite, altered olivine (?).

Rock: Angular to subrounded grains of basalt.

The mineral and rock grains are angular to subrounded and almost all are surrounded by a thin coat of calcium carbonate. The cement of the sandstone is calcium carbonate and clay.

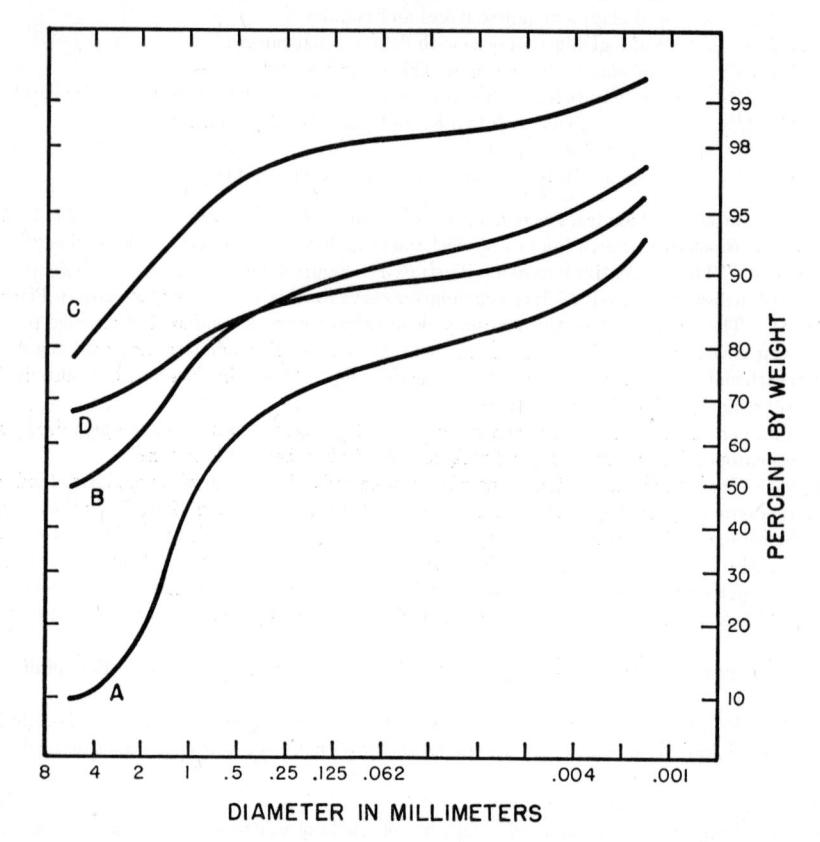

FIGURE 9.—*Cumulative frequency curves from analyses of bottom of core MP 27-2* (D) 4–6 inches from bottom; (C) 0–4 inches from bottom; (B) core catcher; (A) core nose

A complete mechanical analysis was made of this sandstone. The cumulative frequency curve is illustrated in Figure 11. Percentages of the sizes by weight (the Wentworth Scale was used) are as follows:

Sand	75.03
Silt	12.47
Clay	12.50
	100.00

Thirty-four per cent of the sandstone is soluble in HCl. In view of the large amount of calcium carbonate present as cement the mechanical analysis can be considered as approximate only; a mechanical analysis after digestion in HCl might have been approximate also owing to the presence of calcium carbonate as detrital pellets.

The detrital minerals of this sandstone are those of weak resistance to decomposition during erosion and transportation. This fact together with the fact that the mineral grains are angular to subrounded

APPENDIX B. LITHOLOGY AND SEDIMENTS

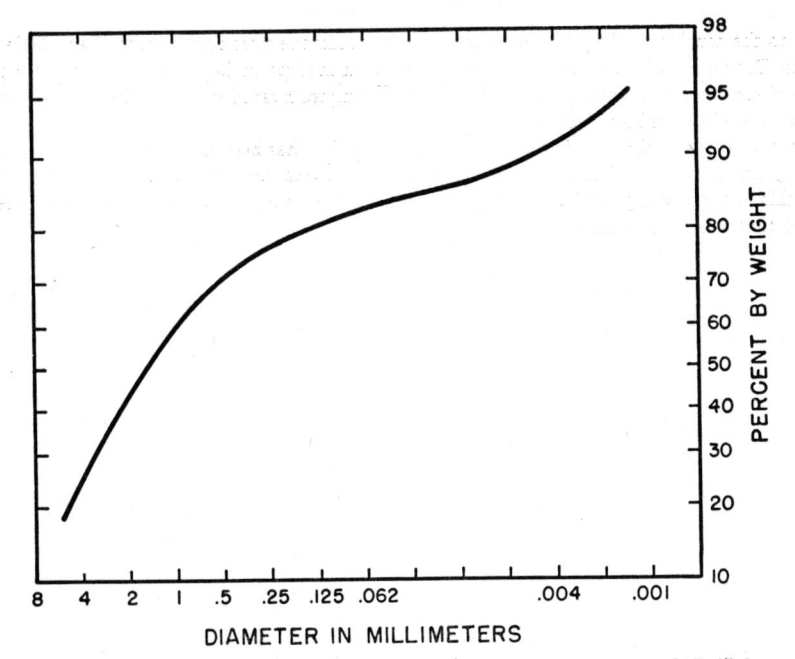

FIGURE 10.—*Cumulative frequency curve from analysis of top of core MP 27-2*

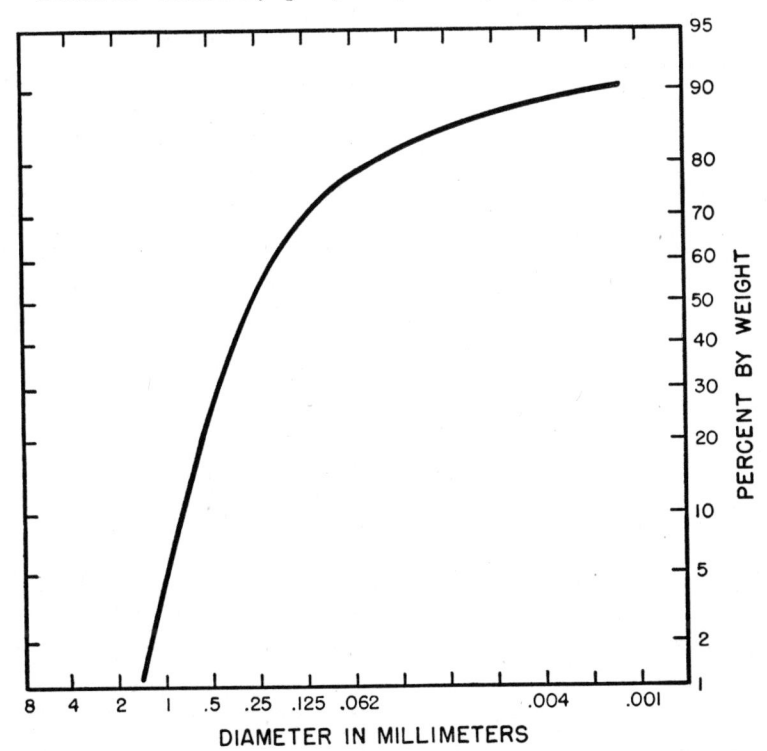

FIGURE 11.—*Cumulative frequency curve from analysis of sandstone from MP 37-A*

leads to the conclusion that the constituents of the sandstone were rapidly eroded and little transported. Dredge haul 37A was made just below the break in slope on Cape Johnson Guyot, the precise place where, after short transportation, material resulting from rapid erosion of the original seamount or island would have been deposited.

MP 37c.—The highly fossiliferous dredge haul at MP 37C has been described under the geology of Cape Johnson Guyot. Thin-section study, chemical analyses, and X-ray powder studies revealed that this material was phosphatized limestone of organic source; many sections of Foraminifera revealed that the original material was a globigerina ooze.

APPENDIX C. SAMPLING EQUIPMENT AND TECHNIQUES

Five different bottom-sampling devices were used in the area of the Mid-Pacific Mountains. The stations and the samplers used are noted in Tables 1–4.

Type of Sampler	No. of times used
Phleger Bottom Sampler (Corer)	20
Chain-bag dredge	17
Piston-type corer	6
Emery-Dietz corer	2
Snapper	1

The snapper used was developed at the Navy Electronics Laboratory. It was based on a snapper developed at the University of California for the sea-floor studies of Shepard, Emery, and others. This snapper is completely described by LaFond and Dietz (1948); essentially it is a spring- and weight-loaded device with jaws that close about a bottom sample when the tripping device strikes the bottom. Three of these snappers were sent down attached to a steel frame at MP 25B with poor results.

The Emery-Dietz corer was used twice on Horizon Guyot. This gravity corer is completely described by Emery and Dietz (1941). Briefly this corer is a pipe shaft with lead weights and small holes that permit the escape of water. A valve housing is screwed into the lower end of the shaft; it contains a rubber stopper which allows water and air to escape upward and prevents loss of the core by suction, by washing through of water during hoisting, or by pressure of overlying water when the corer leaves the surface to be hoisted aboard. A standard pipe core tube is clamped up against a shoulder of the valve housing. Standard lengths of the core barrel are 5, 10, 15, and 20 feet. A core nose and core retainer are screwed to the lower end of the core tube.

The Emery-Dietz corer is lowered to within 30 feet of the bottom at which point most of the winch-brake pressure is removed, and the corer runs freely and plunges into the bottom.

The Emery-Dietz corer has been very successfully used off the coast of California where mud cores have been taken up to 16 feet, 9 inches in length. Emery and Dietz report (1941, p. 1699) after extensive experiments that in this type of gravity corer the length of the core ranges from 40 to 70 per cent of the depth of penetration of the core barrel, although elements of all material penetrated are represented in the core.

On the Mid-Pacific Expedition the new piston-type corer developed by Kullenberg and modified by Mr. J. D. Frautschy of Scripps Institution was used to obtain long cores because this corer was believed most likely to give a true "undisturbed" core of the bottom. Shepard (1948, p. 24) briefly describes this type of instrument. A piston is attached to the end of the cable from the ship and fitted into the core barrel. When the trigger weight strikes the bottom ahead of the core barrel it actuates an arm which allows the weighted core barrel to fall freely around the piston. As the piston slides up in the core barrel, hydrostatic pressure is created which aids the flow of sediment into the core barrel. Pettersson (1954) reports cores up to 65 feet. The longest core taken in the area of the Mid-Pacific Mountains was a 12-foot red-clay core in the deep-sea bottom north of the main ridge (MP 38). The Kullenberg-type piston corer may not give an undisturbed core of the very uppermost layers; to obtain a core of these uppermost layers a Phleger Bottom Sampler was always attached to the trip arm of the piston corer.

The Phleger Bottom Sampler developed by Dr. F. B. Phleger (1951) is a modification of the Ekman gravity coring tube, a pipe with weights and a valve at the top similar to the valve in the Emery-Dietz corer. Inside the core barrel is a plastic tube held in place by the core nose. A core catcher inside the core nose prevents loss of the core from the bottom end. The Phleger corer is operated as is the Emery-Dietz corer. The Phleger corer was used so often (20 times) because of its relative lightness, speed of operating a small winch and light ($3/8$-inch) steel line, and because the use of the plastic liner permitted the core and the water above the core to be corked within the tube and stored without damage. This gravity-type corer is undoubtedly affected by the same factors as is the Emery-Dietz corer and can be expected to take a core whose length is 40–70 per cent of the total depth of penetration.

The major portion of the material on which this paper is based was dredged from the top and sides of the guyots with the chain-bag dredge developed by the Navy Electronics Laboratory. This rugged dredge was developed for deep-sea work. It has a rectangular (36 by 14 inches) boxlike opening made of one-half-inch steel to which is attached a bag made up of one-quarter-inch galvanized iron link chain wired together. A fish-net bag was secured inside the chain bag. Only the larger rocks, nodules, and fossil material were retained inside these wide-meshed bags. To catch the finer material three pipe dredges (short lengths of 2½-inch pipe with canvas bags attached) were trailed behind the large chain-bag dredge. At every station at which these small-pipe dredges were used they came up with the canvas bags solidly packed with globigerina ooze. A "weak link" of steel cable was attached to the steel bridle of the chain-bag dredge and the half-inch steel cable from the ship was attached to the bottom of the bag so that if the dredge became hung up in the rocks on the bottom the "weak link" would break and the dredge would be turned over and pulled out.

In dredging operations on the guyots the ship lay to, and the dredge was lowered on the half-inch steel cable until it was just above the bottom, (as indicated by echo sounder). The HORIZON then moved slowly ahead (at about 2–3 knots) while the cable was payed out. An accumulator which registered the pull on the wire in tons was watched closely to determine when the dredge touched bottom. The accumulator hand will jump several tons as the dredge alternately slides or is caught on the bottom. Best dredging is done up slope with the ship drifting. During the time the dredge was on the bottom (about an hour) the accumulator was watched carefully. When heavier strains approaching the elastic limit of the wire, about half the breaking strain, appeared, the winch brake was slipped to avoid damaging the cable. At one station the HORIZON moved slowly in a complete circle before freeing the dredge which had become caught on the bottom. Only one dredge was lost (MP 25B) out of 17 dredge hauls; no sample was obtained at two stations.

The amount of extra cable payed out is a function of the depth. An analysis of the amount of cable out during the dredge hauls reveals that the average haul had around 25 per cent more wire payed out than the indicated echo-sounder depth which was usually in the range 900–1200 fathoms

APPENDIX D. ECHO SOUNDING

Proper sounding of the sea in order to obtain depths is the most important technique in submarine geology. Accurate soundings are the basis of all bathymetric charts from which studies of the geology of the sea floor are made. Early soundings with rope were highly inaccurate; wire soundings are accurate but time consuming. The development of the echo sounder, by which depth is determined from the time a sound impulse takes to travel to the bottom and back to the surface, revolutionized submarine geology. It is not necessary here to review the history and instruments of deep-sea sounding because such summaries are readily available (Adams, 1942; Shepard, 1948; Kuenen, 1950). Pertinent observations, however, should be made as to the accuracy of echo sounding in view of the fact that many of the results of this paper are based on such soundings.

Small errors in echo sounding, some owing to tide or instrumental difficulties, can be safely disregarded in deep water. The principal error in uncorrected echo soundings is due to variations in the velocity of sound in sea water from the standard velocity (usually 4800 feet per second) to which the fathometer is adjusted. The former velocity depends on three variables: temperature, pressure, and salinity. Because temperature and salinity vary both vertically and horizontally it is necessary to know these factors in order to determine the true velocity of sound at any given location.

The nearest station to the Mid-Pacific guyots at which temperature and salinities have been measured is Station 101 of Cruise VII of the CARNEGIE (Fleming *et al.*, 1945). Temperatures and salinities at this station were extrapolated (by Fleming *et al.*) below 1000 m. These two factors will vary not only from station to station but also at different times at the same station. In the mid-latitudes, however, these fluctuations are normally small, and for the Central Pacific will cause changes of only a few fathoms from one station to the other. The following table (based on Carnegie Station 101) has been checked with several other stations in the Hawaiian area, and the variation from one station to another is commonly less than 2 fathoms at any given depth.

Echo depth (fathoms)	True depth (fathoms)
500	512
1000	1019
2000	2045
2400	2462
2800	2881

Because the fathometer used in obtaining the Mid-Pacific soundings was set for a water velocity of 4800 feet per second, all the soundings obtained were shallower than the true depth (as indicated above). The U. S. Navy Hydrographic Office reports uncorrected soundings on their charts because these are the soundings taken by ships at sea and the ones most useful to a ship's navigator. The various charts of this paper were prepared for the Navy and hence are based on the uncorrected echo-sounding depths. Corrections are minor at the depths of the guyots and on a chart contoured at 200 fathoms.

Modern echo sounders send out an inaudible, high-frequency sound impulse in the form of a cone. The strongest echo will return from a point on the bottom which is tangent to the sound-wave front and shallower than the bottom directly under the ship. As a result the slopes derived from soundings uncorrected for side slopes are not exactly the true slopes on the bottom. In every case the echo-sounder profile is less steep than is the true slope.

Assuming a 30° cone of which the half angle is 15° (thought to be true of the NMC-2 Fathometer), one can see from Figure 12 that a slope recorded as 22° is in truth about 23°. Below 16° there is no correction. Side slopes reported in this paper use uncorrected soundings as is Navy practice and are uncorrected. On most of the slopes reported these corrections would be minor. The reader, should he desire, may readily apply these corrections by reference to Figure 12.

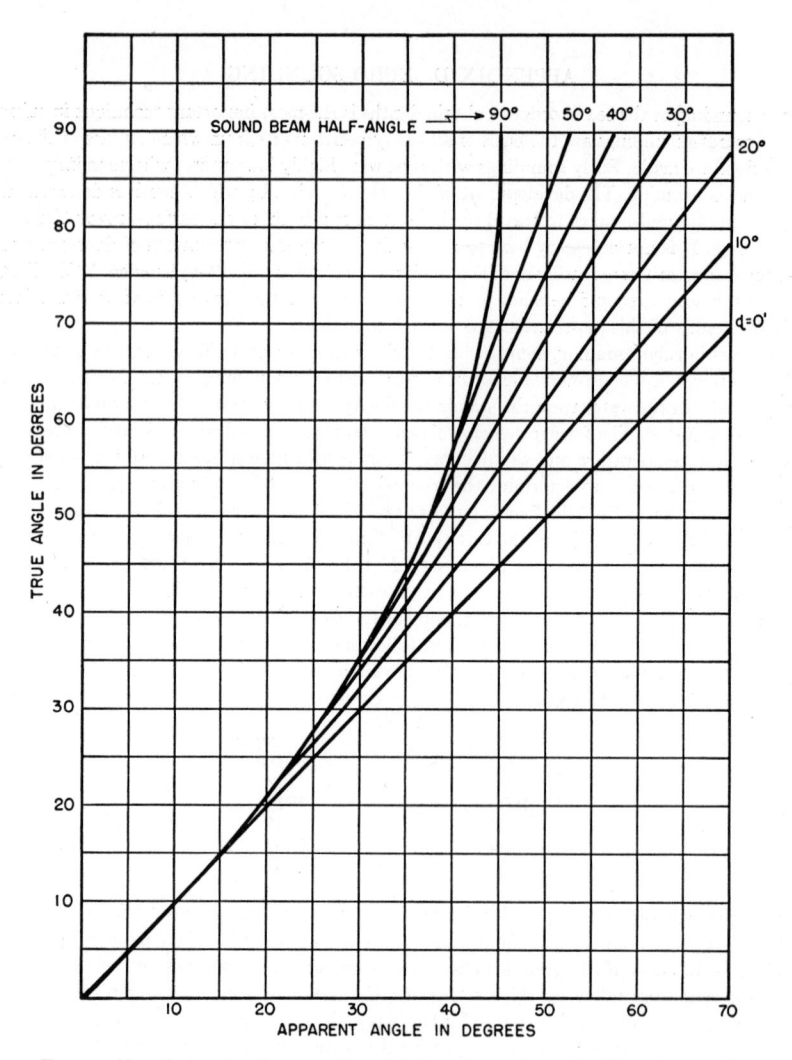

FIGURE 12.—*Curves for the correction of bottom slopes determined by echo sounder* (By R. W. Raitt).

APPENDIX E. LOCATION AT SEA

One of the problems of deep-sea geology is the difficulty of determining true geographical positions. Out of sight of land and as far from land as the guyot area of the Mid-Pacific there are three methods of determining ship position. These are: (1) astronomical fixes, (2) dead reckoning (determination of position by the direction and speed of the ship from a known location) and, (3) Loran.

Loran is a radio location device. The broadcasted pulses and time markers are displayed on the screen of a cathode-ray oscilloscope. Pulses are matched by the use of auxiliary equipment and time-marker pips are counted to obtain the time delay from which location may be derived (Frautschy, unpublished manuscript). Shepard (1948) reports the accuracy of Loran as not more than 15 miles of errors at a distance of 1500 miles. In view of the distance from (and location of) Loran stations this method was of little or no value in Mid-Pac position determinations.

Star fixes are at present the most accurate method of determining geographical positions far from land and was the method used in determining accurate positions on the Mid-Pacific Expedition. James L. Faughan, master of the HORIZON, considered that his navigational fixes by star sights under optimum conditions were accurate within 0.5 of a nautical mile. From one star sight the positions to the next star sight were determined by dead reckoning. After a second star-sight fix has been determined the ship's track between the sights was corrected and drawn up as the "ship's corrected track." The corrected positions between star sights are usually accurate within 0.5–2 nautical miles. The charts and positions reported in this paper were based on the ship's corrected track and can be considered as accurate within the stated limits.

REFERENCES CITED

ADAMS, K. T. (1942) *Hydrographic Manual*, U. S. Dept. Commerce, Coast and Geod. Survey Special Pub. No. 143, Rev. ed.
BARRELL, J. (1917) *Rhythms and measurements of geologic time*, Bull. Geol. Soc. Am., v. 28, p. 776–785.
BARTH, T. F. W. (1931) *Mineralogical petrography of Pacific lavas*, Am. Jour. Sci., v. 21, p. 377, 491.
BELKNAP, G. E. (1874) *Deep-sea soundings in the North Pacific Ocean obtained in the U.S.S. Tuscarora*, U. S. Navy Hydro. Off. No. 54.
BETZ, F., AND HESS, H. H. (1942) *The floor of the North Pacific Ocean*, Geogr. Rev., v. 32, p. 99–116.
BLANKENHORN, M. (1927) *Die fossilen Gastropoden und Scaphopoden der Kreide von Syrien-Palestina*, Paleontographica, v. 69, p. 111–186.
BRAMLETTE, M. N. (1946) *The Monterey formation of California and the origin of its siliceous rocks*, U. S. Geol. Survey Prof. Paper 212.
BRYAN, G. S. (1940) *A bathymetric chart of the North Pacific Ocean*, Assn. d'Oceanogr. Phys., Union Geodes. et Geophys. Int., Publ. Sci. No. 8, p. 78–80.
CAMPBELL, D. H. (1919) *The derivation of the flora of Hawaii*, Lel. Stanford Univ. Publ.; *The Australian element in the Hawaiian flora*, Proc. 3rd Pan-Pac. Sci. Congr., v. 1 (1929), p. 938–946.
CARSOLA, A. J., AND DIETZ, R. S. (1952) *Submarine geology of two flat-topped northeast Pacific seamounts*, Am. Jour. Sci., v. 250, p. 481–497.
CHAMBERLIN, T. C. (1897) *The method of multiple working hypotheses*, Jour. Geol., v. 5, p. 837–848.
CHUBB, G. J. (1934) *The structure of the Pacific Basin*, Geol. Mag., v. 71, p. 289–301.
CLARKE, F. W. (1920 and 1924) *The data of geochemistry*, Bull. U. S. Geol. Survey Nos. 695 and 770.
CLARK, B. L. (1945) *Problems of speciation and correlation as applied to mollusks of the marine Cenozoic*, Jour. Paleon. v. 19, p. 158–172.
COOKE, C. M. (1923) *The distribution of Hawaiian land-snails*, Proc. Pan-Pac. Sci. Congr., v. 2, p. 1545–1548.
COOKE, C. W. (1946) *Comanche echinoids*, Jour. Paleon., v. 20, p. 193–237.
COTTON, C. A. (1944) *Volcanoes as Landscape Forms*, Wellington, N. Z., Whitcombe and Tombs.
CUSHMAN, J. A. (1946) *Upper Cretaceous Foraminifera of the Gulf Coastal Region of the United States and adjacent areas*, Geol. Survey Prof. Paper 206.
DALY, R. A. (1910) *Pleistocene glaciation and the coral reef problem*, Am. Jour. Sci., Ser. 4, v. 30, p. 297–308.
——— (1916) *Petrography of the Pacific Islands*, Bull. Goel. Soc. Am., v. 27, p. 325–344.
——— (1933) *Igneous rocks and the depths of the earth*, N. Y., McGraw-Hill.
——— (1942) *The Floor of the Ocean*, Univ. of N. Carolina Press.
DAMES, W. (1877) *Die Echiniden der vicentinischen und veronesischen Tertiaerablagerungen*, Paleontographica, v. 25, p. 1–100.
DARWIN, C. (1837) *On certain areas of elevation and subsidence in the Pacific and Indian Oceans, as deduced from the study of coral formations*, Proc. Geol. Soc. London, v. 2, p. 552–554.
——— (1842) *The Structure and Distribution of Coral Reefs*, 3rd ed., 1889, N. Y., Appleton.
DAVIES, A. M. (1934) *Tertiary Faunas*, London, Thomas Murby and Co.
DAVIS, W. M. (1919) *The geological aspects of the coral-reef problem*, Sci. Progress, v. 3, p. 420–444.
——— (1928) *The Coral Reef Problem*, Am. Geogr. Soc. Special Publ. No. 9.
DEHORNE, Y. (1920) *Les stromatoporoids des Terrains Secondaires*, Memoire pour servir l'explication de la carte geologique de France, 170 p.
DES MOULINS, C. (1835–37) *Etudes sur les echinides*, Actes Soc. Linnéenne de Bordeaux, v. 7, p. 167–227, 315–432; v. 9, p. 45–364.
DIETZ, R. S., AND MENARD, H. W. (1951) *Origin of abrupt change in slope at continental shelf margin*, Bull. Am. Assoc. Petrol. Geol., v. 35, p. 1994–2016.
——— (1953) *Hawaiian swell, deep, and arch, and subsidence of the Hawaiian Islands*, Jour. Geol., v. 61, p. 99–113.
DIETZ, R. S., EMERY, K. O., AND SHEPARD, F. P. (1942) *Phosphorite deposits on the sea floor off Southern California*, Bull. Geol. Soc. Am., v. 53, p. 815–848.

DOBRIN, M. B., PERKINS, B., JR., AND SNAVELY, B. L. (1949) *Subsurface constitution of Bikini Atoll as indicated by a seismic refraction survey,* Bull. Geol. Soc. Am., v. 60, p. 807–828.

D'ORBIGNY, A. (1853–60) *Paleon. Francaise, Terrane Cret.,* v. 1–8, Paris, G. Masson.

DOUVILLE, H. (1898) *Sur les rudistes due Gault Superieur du Portugal,* Bull. Soc. Geol. Fr., 3rd ser., v. 26, p. 140–151.

EMERY, K. O. (1948) *Submarine geology of Ranger Bank, Mexico,* Bull. Geol. Soc. Am., v. 32, p. 790–805.

EMERY, K. O., AND DIETZ, R. S. (1941) *Gravity coring instruments and mechanics of sediment coring,* Bull. Geol. Soc. Am., v. 52, p. 1685–1714.

———— (1950) *Submarine phosphorite deposits off California and Mexico,* Calif. Jour. Mines and Geology, v. 46, No. 1.

EMERY, K. O., TRACEY, J. I., JR., AND LADD, H. S. (1954) *Geology of Bikini and nearby atolls,* U. S. Geol. Survey Prof. Paper 260-A.

EMMONS, W. H. (1940) *The Principles of Economic Geology,* N. Y., McGraw-Hill.

FELIX, J. (1891) *Versteinerungen aus der mexicanischen Jura- und Kreideformation,* Paleontographica, v. 37, p. 140–199.

FLEMING, J. A., SVERDRUP, H. U., ENNIS, C. C., SEATON, S. L., AND HENDRIX, W. C. (1945) *Scientific results of Cruise VII of the Carnegie—,* Oceanography I-B, Carnegie Inst. of Wash. Pub. 545.

FREEMAN, O. W. (1951) *Geography of the Pacific,* N. Y., John Wiley and Sons.

GARDINER, J. S. (1931) *Coral Reefs and Atolls,* N. Y., Macmillan Co.

GOLDBERG, E. D. (1954) *Marine geochemistry 1. chemical scavengers of the sea,* Jour. Geol., v. 62, p. 249–265.

GREGORY, J. W. (1930) *The geological history of the Pacific Ocean,* Quart. Jour. Geol. Soc., v. 86, lxxii-cxxxxvi.

GRIGGS, D. (1939) *A theory of mountain building,* Am. Jour. Sci., v. 237, p. 611–650.

GUNN, R. (1943) *A quantitative study of isobaric equilibrium and gravity anomalies in the Hawaiian Islands,* Franklin Inst. Jour., v. 236, p. 373–390.

———— (1947) *Quantitative aspects of juxtaposed ocean deeps, mountain chains, and volcanic ranges,* Geophysics, v. 12, p. 238–255.

———— (1949) *Isostasy extended,* Jour. Geol., v. 57, p. 263–280.

GUTENBERG, B., AND RICHTER, C. F. (1941) *Seismicity of the earth,* Geol. Soc. Am., Special Paper No. 34.

HACKEMESSER, M. (1936) *Eine kretazische Korallenfauna aus Mittel-Griechenland,* Paleontographica, v. 84 (A), p. 1–97.

HAMILTON, E. L. (1953) *Upper Cretaceous, Tertiary, and Recent planktonic Foraminifera from Mid-Pacific flat-topped seamounts,* Jour. Paleon., v. 27, p. 204–237.

HANZAWA, S. (1940) *Micropaleontological studies of drill cores from a deep well in Kita-Daito-Zima* (North Borodino Island), Jubilee Pub. in Commemoration of Prof. H. Yabe's 60th Birthday, v. 2, p. 755–802.

HAUG, E. (1909) *Traite de Geologie,* Paris, Armand Colin.

HERTLEIN, L. G. (1937) *A note on some species of marine mollusca occurring in both Polynesia and Western America,* Am. Phil. Soc. Proc., v. 78, p. 310–312.

HESS, H. H. (1946) *Drowned ancient islands of the Pacific Basin,* Am. Jour. Sci., v. 244, p. 772–791.

———— (1948) *Major structural features of the Western North Pacific, an interpretation of H. O. 5485, Bathymetric Chart, Korea to New Guinea,* Bull. Geol. Soc. Am., v. 59, p. 417–446.

———— (1954) *Geological hypotheses and the earth's crust under the oceans,* Proc. Royal Soc. London, v. 222, p. 341–348.

HINDE, G. J. (1904) *Report on Funafuti Atoll boring,* in Sollas, W. J., et al., Rept. of Coral Reef Comm. of the Royal Soc., London, Harrison and Sons.

———— (1944) *Mountain growth, a study of the southwest Pacific region,* Proc. Am. Phil. Soc., v. 88, p. 221–268.

HOFFMEISTER, J. E., AND LADD, H. S. (1944) *The Antecedent-Platform Theory,* Jour. Geol., v. 52, p. 388–402.

HOFFMEISTER, J. E., LADD, H. S., AND ALLING, H. L. (1929) *Falcon Island,* Am. Jour. Sci., v. 218, p. 461–471.
INTERNATIONAL HYDROGRAPHIC BUREAU (1949) *General bathymetric chart of the oceans,* Internat. Hydro. Bureau, Special Pub. No. 30, Pt. A II.
JOHNSON, D. W. (1933) *Role of analysis in scientific investigation,* Bull. Geol. Soc. Am., v. 44, p. 461–494.
KIRK, M. V., AND MACINTYRE, J. R. (1951) *Cretaceous deposits of the Punta San Isidro area, Baja California,* (abstract), Bull. Geol. Soc. Am., v. 62, p. 1505.
KUENEN, P. H. (1933) *Geology of coral reefs, in* Snellius Expedition 1929–30, v. 5, pt. 2., Leyden, Brill.
────── (1937) *On the total amount of sedimentation in the deep sea,* Am. Jour. Sci., v. 34, p. 457–468.
────── (1950) *Marine Geology,* N. Y., John Wiley and Sons.
KÜHN, O. (1933) *Das becken von Isfahan-Saidabad und seine altmiocane Korallen fauna,* Paleontographica, v. 79 A, p. 143–218.
KULP, J. L. (1951) *Origin of the hydrosphere,* Bull. Geol. Soc. Am., v. 62, p. 326–329.
LADD, H. S. (1950) *Recent Reefs,* Bull. Am. Assoc. Petrol. Geol., v. 34, p. 203–214.
LADD, H. S., AND HOFFMEISTER, J. E. (1936) *A criticism of the glacial control theory,* Jour. Geol., v. 44, p. 74–92.
LADD, H. S., INGERSON, E., TOWNSEND, R. C., RUSSELL, M., AND STEPHENSON, H. K. (1953) *Drilling on Eniwetok Atoll, Marshall Islands,* Bull. Am. Assoc. Petrol. Geol., v. 37, p. 2257–2280.
LADD, H. S., AND TRACEY, J. I., JR. (1949) *The problem of coral reefs,* Scientific Monthly, v. 69, p. 297–305.
LADD, H. S., TRACEY, J. I., JR., AND LILL, G. G. (1948) *Drilling on Bikini Atoll, Marshall Islands,* Science, v. 107, p. 51–55.
LADD, H. S., TRACEY, J. I., JR., WELLS, J. W., AND EMERY, K. O. (1950) *Organic growth and sedimentation on an atoll,* Jour. Geol., v. 58, p. 410–425.
LAFOND, E. C., AND DIETZ, R. S. (1948) *New snapper-type sea floor sediment sampler,* Jour. Sed. Petrol., v. 18, p. 34–37.
LAWSON, A. C. (1895) *Sketch of the geology of the San Francisco peninsula,* U. S. Geol. Survey 15th Ann. Rept., p. 405–476.
LOBECK, A. K. (1939) *Geomorphology,* N. Y., McGraw-Hill.
LULL, R. S. (1929) *Organic Evolution,* N. Y., Macmillan Co.
MACDONALD, G. A. (1949a) *Petrography of the Island of Hawaii,* U. S. Geol. Survey Prof. Paper No. 214-D.
────── (1949b) *Hawaiian petrographic province,* Bull. Geol. Soc. Am., v. 60, p. 1541–1596.
MATHEWS, W. H. (1947) *"Tuyas", flat-topped volcanoes in N. British Columbia,* Am. Jour. Sci., v. 245, p. 560–570.
MATTHEW, W. D. (1915) *Climate and Evolution,* Ann. N. Y. Acad. Sci., v. 25, p. 171–318.
MATSUYAMA, M. (1918) *Determinations of the second derivatives of the gravitational potential on Jaluit Atoll,* College of Sci., Kyoto, Japan, Mem. 3, p. 17–68.
MENARD, H. W. (1952) *Deep ripple marks in the sea,* Jour. Sed. Petrol., v. 22, p. 3–9.
────── (1955) *Deformation of the northeastern Pacific Basin and the West Coast of North America,* Bull. Geol. Soc. Am., v. 66, p. 1149–1198.
MENARD, H. W., AND DIETZ, R. S. (1951) *Submarine geology of the Gulf of Alaska,* Bull. Geol. Soc. Am., v. 62, p. 1263–1286.
MORTENSEN, TH. (1948) *A monograph of the Echinoids,* C. A. Reitzel, Copenhagen, v. 4, pt. 1.
MULLER, S. W., AND SCHENCK, H. G. (1943) *Standard of the Cretaceous System,* Bull. Am. Assoc. Petrol. Geol., v. 27, p. 262–278.
MURRAY, H. W. (1941) *Submarine mountains in the Gulf of Alaska,* Bull. Geol. Soc. Am., v. 52, p. 333–362.
MURRAY, J. (1877) *On the distribution of volcanic debris over the floor of the ocean,* Proc. Royal Soc. Edin., v. 9, p. 247.
MURRAY, J., AND RENARD, A. F. (1891) *Report on the deep-sea deposits based on the specimens collected during the voyage of HMS Challenger in the years 1872 to 1876,* London, Longmans, p. 1–525.

MURRAY, J., AND LEE, G. V. (1909) *The depth and marine deposits of the Pacific*, Mem. Mus. Comp. Zool., v. 38, p. 1–169.

NEAVERSON, E. (1928) *Stratigraphical Paleontology*, London, Macmillan.

OLIVER, J. E., EWING, MAURICE, AND PRESS, FRANK (1955) *Crustal structure and surface wave dispersion*, Bull. Geol. Soc. Am., v. 66, p. 913–946.

PALMER, R. H. (1928) *The rudistids of southern Mexico*, Occasional Papers of the Calif. Acad. Sci., No. 14.

PAQUIER, V. (1903) *Les rudistes Urgoniens*, Mem. Soc. Geol. Fr., Paleon. No. 29, p. 1–102.

PARK, C. F., JR. (1946) *The spilite and manganese problems of the Olympic Peninsula, Washington*, Am. Jour. Sci., v. 244, p. 305–323.

PARONA, C. F. (1932) *Di alcuni idrozoi del Giurassico e Cretacico in Italia*, Mem. della Royal Acad. Sci. Torino, Ser II, v. 67, No. 7.

PATTERSON, C. C., GOLDBERG, E. D., AND INGHRAM, M. G. (1953) *Isotopic compositions of Quaternary leads from the Pacific Ocean*, Bull. Geol. Soc. Am., v. 64, p. 1387–1388.

PETTERSSON, H. (1954) *The ocean floor*, New Haven, Yale Univ. Press.

PHLEGER, F. B. (1951) *Ecology of Foraminifera, northwest Gulf of Mexico, Part I Foraminifera Distribution*, Geol. Soc. Am. Mem. 46.

PILSBRY, H. H. (1899) *The genesis of Mid-Pacific faunas*, Proc. Acad. Nat. Sci., Phila., (Nov. 1900), p. 576, 578, 581.

RAITT, R. W. (1952) *The 1950 seismic refraction studies of Bikini and Kwajalein Atolls and Sylvania Guyot*, Scripps Inst. Oceanography Reference 52-38, p. 1–25.

———— (1953) *Reflection and refraction of explosive waves by the sea bottom*, Scripps Inst. Oceanography Quarterly Progress Rept., Reference 53-22, p. 1–3.

REVELLE, R. R. (1944) *Marine bottom samples collected in the Pacific by the Carnegie on its seventh cruise*, Carnegie Inst. of Wash., Pub. 556, p. 1–180.

ROY, C. J. (1945) *Silica in natural waters*, Am. Jour. Sci., v. 243, p. 393–403.

RUBEY, W. W. (1929) *Origin of the siliceous Mowry shale of the Black Hills region*, U. S. Geol. Survey Prof. Paper 154.

———— (1951) *Geologic history of sea water; an attempt to state the problem*, Bull. Geol. Soc. Am., v. 62, p. 1111–1148.

SCHUCHERT, C. (1932) *The periodicity of oceanic spreading, mountain making, and paleogeography*, Bull. Nat. Res. Coun. 85, p. 537–561.

SHEPARD, F. P. (1948) *Submarine Geology*, N. Y., Harper and Bros.

SHIMER, H. W., AND SHROCK, R. R. (1944) *Index Fossils of North America*, N. Y., John Wiley and Sons; 2nd printing, 1947.

SIMPSON, G. G. (1940) *Mammals and land bridges*, Jour. Wash. Acad. Sci., v. 30, p. 136–163.

SKOTTSBERG, C. (1926) *Remarks on the relative independency of Pacific floras*, Proc. 3rd Pan Pac. Sci. Cong., v. 1 (1929), p. 917.

STEARNS, H. T. (1946a) *Geology of the Hawaiian Islands*, Hawaii Div. of Hydro., Bull. 8.

———— (1946b) *An integration of coral-reef hypotheses*, Am. Jour. Sci., v. 244, p. 245–262.

STEINER, A. (1932) *Contribution a l'etude des Stromatopores Secondaires*, Mem. Soc. Vaudrise Sci. Nat., v. 4, p. 105–225.

SUESS, E. (1893) *"Are great ocean depths permanent?"*, Nat. Sci., v. 2, p. 183.

SVERDRUP, H. U., JOHNSON, M. W., AND FLEMING, R. H. (1946) *The Oceans*, N. Y., Prentice-Hall, Inc.

SWINNERTON, H. H. (1947) *Outlines of Paleontology*, 3rd ed., London, Edward Arnold and Co.

TALIAFERRO, N. L. (1933) *The relation of vulcanism to diatomaceous and associated siliceous sediments*, Univ. Calif. Pub., Dept. Geol. Sci. Bull., v. 23, p. 1–55.

UMBGROVE, J. H. F. (1947) *The Pulse of the Earth*, The Hague, Nijhoff.

VAUGHAN, T. W. (1919a) *Coral and the formation of coral reefs*, Ann. Rept. Smith. Inst. for 1917, p. 189–276.

———— (1919b) *Fossil corals from Central America*, U. S. Nat. Mus. Bull. 103.

VAUGHAN, T. W., AND WELLS, J. W. (1943) *Revision of the suborders, families, and genera of the Scleractinia*, Geol. Soc. Am Special Paper No. 44.

REFERENCES CITED

VENING MEINESZ, F. A. (1934) *Gravity expeditions at sea*, v. 2, Netherlands Geodetic Commission.

——— (1941) *Gravity over the Hawaiian Archipelago and over the Madeira area*, Proc. Kon. Akad. v. Wettensch., Amsterdam, v. 45, p. 120–125.

——— (1944) *De verdeling van continenten en oceanen over het aardoppervlak*, Versl. Kon. Akad. v. Wetensch., Amsterdam, Afd. Nat., v. 53, p. 151–159.

——— (1948a) *Gravity expeditions at sea, 1923–1938*, v. 4, Netherlands Geodetic Commission.

——— (1948b) *Major tectonic phenomena and the hypothesis of convection currents in the earth*, Quart. Jour. Geol. Soc. London, v. 103, p. 191–207.

WALLACE, A. R. (1880) *Island Life*, London, Macmillan.

WASHINGTON, H. S. (1923) *Petrology of the Hawaiian Islands: I–IV*, Am. Jour. Sci., v. 5, p. 465–502; v. 6, p. 100–126, 338–367, 409–423.

———, AND KEYES, M. G. (1926) *Petrology of the Hawaiian Islands: V*, Am. Jour. Sci., v. 12, p. 336–352.

——— (1927) *Rocks of the Galapagos Islands (and of Hawaii)*, Wash. Acad. Sci. Jour., v. 17, p. 538–543.

——— (1928) *Petrology of the Hawaiian Islands: VI*, Am. Jour. Sci., v. 15, p. 199–220.

WEGENER, A. L. (1922) *The Origin of the Continents and Oceans*, (Trans. by J. G. A. Skerl), N. Y., E. P. Dutton and Co.

WELLS, J. W. (1932) *Corals of the Trinity Group of the Comanchean of Central Texas*, Jour. Paleon., v. 6, p. 225–256.

——— (1933) *Corals of the Cretaceous of the Atlantic and Gulf coastal plains and western interior of the United States*, Bull. Am. Paleon., v. 18, No. 67.

——— (1934a) *Eocene corals from Cuba*, Pt. 1, Bull. Am. Paleon., v. 20, No. 70 B.

——— (1934b) *Some fossil corals from the West Indies*, U. S. Nat. Mus. Proc., v. 83, p. 71–110.

——— (1934c) *A New Species of Stromatoporoid from the Buda Limestone of Central Texas*, Jour. Paleon., v. 8, p. 169.

——— (1944) *Cretaceous, Tertiary and Recent corals, a sponge and an algae from Venezuela*, Jour. Paleon., v. 18, p. 429–447.

WHARTON, A. J. L. (1897) *Foundations of coral atolls*, Nature v. 55, p. 390–393.

WILLIAMS, H. (1941) *Calderas and their origin*, Univ. of Calif. Pub., Dept. of Geol., v. 25, No. 6.

WOODS, H. (1950) *Paleontology, Invertebrate*, Cambridge Univ. Press.

YABE, H., AND SUGIYAMA, T. (1935) *Jurassic stromatoporoids from Japan*, Sci. Rept. of Tohoku Imp. Univ., Sendai, Japan, ser. 2, v. 14, No. 2 B, p. 136–192.

INDEX

Actinostroma, 22, 27
   pacifica, 61, 73
Adams, K. T., 83
ALBATROSS, HMS, 34
Algae, coralline, 18, 28, 29
Anderson, T., 42
Andesite Line, 32, 47
Anthozoa (*See* Coral)
Astrocoenia, 22, 23
   dietzi, 57, 71
   revellei, 57, 71

Barrell, J., 33
Barth, T. F. W., 32
Basalt
   detailed descriptions, 75–80
   erosional debris, 32, 33
   occurrence on guyots, 5, 9, 10, 11, 12, 14, 18, 20, 40
   summary of samples collected, 32
   under Pacific basin, 32
Belknap, G. E., 4
Betz, F., 43
Bikini Atoll, 34, 48
Blankenhorn, M., 28
Brachyseris, 22, 23, 58
   montemarina, 58, 72
   wellsi, 58, 72
Bramlette, M. N., 35
BRITANNIA, HMS, 34
Bryan, G. S., 4

Calcareous ooze, 9, 18, 20, 33, 34
Campbell, D. H., 51
Cape Johnson Guyot, 9, 18–22, 23, 26, 27, 28, 29, 32, 33, 34, 35, 36, 37, 39, 40, 42, 53, 58, 59, 61, 64, 65, 66, 71, 80
   fossil faunas on, 20–22
   morphology of, 18–20
   naming of, 18
   samples from, 20–22
Caprina, 22, 23, 26
   choffati, 26
   fragments, 66
   mediopacifica, 66, 74
   mulleri, 65, 74
Cardita, 22
   sp., 66
Carsola, A. J., 3, 32, 52
Cerithium, 22, 28, 64
   sp., 64, 72

Chain-bag dredge, 81, 82
CHALLENGER, HMS, v, 29, 34, 36
Chamberlin, T. C., 38
Chubb, G. J., 43
Clark, B. L., 50, 51
Clarke, F. W., 37
Convection currents, subcrustal, 47–48
Cooke, C. M., 51
Cooke, C. W., 28, 67, 68
Coquina, 18, 33, 77
Coral
   affinities of fossil, 23–24
   age of fossil, 21, 22, 23, 24
   conditions of fossilization, 25
   death of original fauna, 25–26
   ecologic conditions, 24, 25, 26
   general discussion, 23–26
   migration of larvae of, 51, 52
   occurrence on guyots, 18, 20, 22, 25, 33, 43
   solitary, 28
   systematic descriptions, 57–61
Coral-Atoll Theories
   Antecedent Platform Theory, 49
   bearing of conclusions on, 48–50
   Daly's Glacial Control Theory, 49
   Darwin's Subsidence Theory, 48–50
Cores at MP 27, 12–14, 32, 76, 77
Core samplers
   Emery-Dietz, 81
   Kullenberg, 12, 20, 81
   Phleger, 10, 12, 20, 81
Coring operations, 5, 9, 10, 12, 15, 18, 20
   detailed sample descriptions, 75–80
   discussion of equipment, 81
Cotton, C. A., 40
Cretaceous
   fossil fauna (foraminiferal), 14, 21, 29–31, 68
   fossil fauna (megafossils), 18, 21–29
      affinities, 23, 52
      ecology, 25
      systematic descriptions, 57–68
Cushman, J. A., 30
Cyathophora, 22, 23, 60

Daly, R. A., 32, 49
Dames, W., 28
Darwin, C., 48, 49, 50, 62
Davies, A. M., 23, 50, 51
Davis, W. M., 26, 48, 49, 50, 51, 63
Dehorne, Y., 27, 62

Des Moulins, C., 67
Diceras, 26
Dietz, R. S., v, 2, 3, 4, 9, 28, 32, 33, 36, 39, 43, 52, 58, 81
Diploastrea, 22, 23
  sp., 60, 72
Dobrin, M. B., 48
d'Orbigny, A., 23, 28, 67
Douville, H., 26
Dredging operations, 5, 9, 10, 12, 18, 20
  description of, 82
  fossils from, 71-75
  hauls from, descriptions, 75-80
Durham, J. W., 28
Dutch Siboga Expedition, 26

Earbone, cetacean, 20, 29, 75
Echinoid (*See Pyrina*)
Echo sounding
  corrections to get true depth, 83
  errors in, 83
  side-slope corrections for, 83, 84
Emery, K. O., 3, 4, 36, 39, 40, 42, 44, 52, 81
Emery-Dietz Corer, 81
Emmons, W. H., 35
Eocene Foraminifera, 9, 11, 30, 31, 68-70
Erben Guyot, 3, 52
Erosion of guyots, 32-33
  (*See also* Truncation of guyots)

Faughan, James L., 85
Faunal migration
  methods of, 50-52
  modification of theories of, 50-52
  on "island stepping-stones", 51-52
Felix, J., 28
Fieberling Guyot, 3, 32, 52
Flat-topped seamounts (*See* Guyots)
Fleming, J. A., 83
Foraminifera
  age, affinities, ecology, 30-31, 34
  Cretaceous (*See* Cretaceous)
  Eocene (*See* Eocene Foraminifera)
  faunal lists of, 68-70
  fossil planktonic, 9, 11, 12, 14, 18, 21, 22, 29-31, 34
  Tertiary, 9, 11, 12, 18, 22, 29, 30, 31, 34, 68-70
Fossil faunas (*See* Paleontology)
Frautschy, J. D., 81
Freeman, O. W., 51
Fuller, R. E., 42

Gardiner, J. S., 28
Gastropods, 20, 22, 28, 33, 63-65, 77

Geologic history of Mid-Pacific Mountains and guyots, 54-55
Geology of the guyots (*See* Guyots)
Geomorphology, 38-44
  of guyots, 39-43
  of Mid-Pacific mountains, 43-44
  truncation of guyots, 41-43
Globigerina ooze
  detailed sample descriptions, 75-80
  fossil, 9, 11, 12, 14, 18, 30, 34
  from TUSCARORA samples, 4
  indurated, 11, 18, 20, 27, 29, 30, 31, 34, **35, 36**
  occurrence of, 33-35
  phosphatized, 21, 35, 36
Goldberg, E. D., 36, 37
Gravity anomalies
  Hawaiian Islands, 46-47
  Mid-Pacific Mountains, 46-47
Gregory, J. W., 51
Griggs, D., 47, 48, 54
Gulf of Alaska
  seamounts and guyots in, 2-3
Gunn, R., 46, 47, 48
Gutenberg, B., 47
Guyot, Arnold, 2
Guyots (flat-topped seamounts)
  distribution of, 2-3
  Erben Guyot, 3
  Fieberling Guyot, 3
  first noted, 2
  geology of, 5-22, 54-55
    Cape Johnson Guyot, 18-22
    Guyot 19171, 9-11
    Guyot 20171, 11-12
    Hess Guyot, 14-18
    Horizon Guyot, 5-9
  geomorphology of, 38-43
  hypotheses as to land forms, 39
  in Gulf of Alaska, 2-3
  in Marshall Islands, 3-4
  naming of, 2
  plans to explore, v
  previous investigations, 2-4
  previous opinions on, 1
  submergence of, 44-48
  truncation of, 41-43
Guyot 19171, 9-11, 28, 32, 33, 34, **70**
  fossil faunas on, 11
  morphology of, 9-10
  naming of, 9
  samples from, 10-11
Guyot 20171, 11-12, 32, 33, 53
  fossil faunas on, 12

INDEX

morphology, 11–12
samples from, 12

Hackemesser, M., 60
Hamilton, E. L., 30, 31, 57
Hanzawa, S., 48
Harvey, C. E., 36
Haug, E., 26
Hawaiian Islands, 42, 43
  faunal affinities of, 51, 52
  isostatic adjustments of, 46–47
  Necker I., 4, 44
  structure of, 43
Hawaiian Swell, 43
Heiskanen, W., 43, 47
Hertlein, L. G., 51
Hess Guyot, 9, 14–18, 22, 23, 27, 28, 29, 31, 32, 33, 34, 35, 39, 40, 42, 53, 57, 58, 59, 60, 61, 62, 63, 71
  fossil faunas on, 18
  morphology of, 14–15
  naming of, 14
  samples from, 15–18
Hess, H. H., 1, 2, 4, 5, 14, 18, 29, 32, 43, 45, 47, 53
Hinde, G. J., 48
Hobbs, W. H., 32
Hoffmeister, J. E., 41, 42, 49
Horizon Guyot, 5–9, 29, 30, 31, 32, 33, 34, 81
  fossil faunas on, 9
  morphology of, 5
  naming of, 5
  samples from, 5, 9
HORIZON, R/V, v, 4, 12, 82

Igneous rocks, 32–33
  detailed descriptions of, 75–80
  occurrence on guyots, summary of, 32
  (*See also* individual guyots and Basalt)
International Hydrographic Bureau (Monaco), 4
Isostatic adjustments, 46–48

Johnson, D. W., 38

Keen, A. Myra, 64, 68
Keyes, M. G., 32
Kirk, M. V., 28
Koto, B., 42
Kuenen, P. H., 32, 34, 39, 44, 45, 46, 47, 83
Kühn, O., 28
Kullenberg piston corer, 12, 20, 81
Kulp, J. L., 45

Ladd, H. S., 24, 26, 33, 41, 42, 48, 49, 50
LaFond, E. C., 81

Lava (*See also* Basalt and Igneous rock)
  flows as material of guyots, 41
  subaqueous flows, 42, 43
Lawson, A. C., 35
Lee, G. V., 34, 35, 36
Limestone
  (*See also* Globigerina ooze, indurated)
  calcareous oozes, 9, 18, 20, 33, 34
  detrital limestone, 18, 35
  phosphatized limestone, 9, 18, 20, 21, 35–36, 54, 61, 75, 76, 77, 80
  silicified limestone, 9, 35, 37, 38, 54, 75, 76
  summary of occurrence, 33–36
Lithology (*See* specific rock type)
Lobeck, A. K., 39
Location at sea, 85
Lophosmilia, 22
  fundimaritima, 60, 71
Lull, R. S., 51

Macdonald, G. A., 32, 42
MacIntyre, J. R., 28
Manganese dioxide
  general discussion of, 36–38
  manganese crusts or coatings, 5, 9, 11, 18, 20, 34, 36, 75–77
  manganese nodules, 5, 9, 11, 12, 18, 20, 35, 36, 38, 75–77
    spectrophotometric analyses of, 37
  occurrence on guyots, 5, 9, 10, 36
  research and theories of formation of, 37, 38
Manganese nodules (*See* Manganese dioxide)
Marshall Islands
  Bikini, 49, 50
  Eniwetok, 49, 50
  guyots and seamounts of, 3, 40, 44
Mathews, W. H., 42
Matsuyama, M., 48
Matthew, W. D., 50
Menard, H. W., 2, 4, 33, 34, 39, 43, 44, 59
Microfossils (*See* Foraminifera)
Microsolena, 22, 23
  sp., 59, 72
Mid-Pacific 1950 Expedition, v, 1, 3, 4, 38, 48, 50
  purpose and itinerary, v
Mid-Pacific Mountains
  age when islands, 29, 41–43
  definition of, 4
  geomorphology of, 38–44
  isostatic adjustments of, 46–48
  physiography of, 4–5, 43–44
  previous expedition in, 4
  rocks of, 32–33
  structure and origin of, 43–44

Mid-Pacific Mountains (Cont'd)
  submergence of, 29, 42–48
  truncation of peaks of, 41–43
Milleporidium, 18, 22, 27
  darwini, 62, 73
  davisi, 62, 73, 75
Molengraaff, G. A. F., 26
Montastrea, 22, 23
  menardi, 59, 71
Mortensen, Th., 68
Muller, S. W., 24, 65
Multiple Working Hypotheses
  Theory of, 38
Murray, H. W., 2
Murray, J., 29, 34, 35, 36, 37

Neaverson, E., 28, 68
Nerinea, 22, 28, 64
  sp., 64

Oceans
  Cretaceous currents, 52
  depth of Cretaceous Pacific, 29
  origin of, 44–45
  thickness of sediments in, 45–46
  volume of, 44–45
Oculininae (Solitary coral), 28, 61, 75
Oliver, J. E., 45
Oxyrhina, 28

Pacific Basin
  oldest rocks known in, 29
  rocks of, 32
  stability of, 47
  subsidence in, 48
Paleontology
  megafossils, 22–29, 57–68
    affinities of, 29, 52
    age of, 23, 24, 26, 29
    assemblages, 22
    coral, 23–26 (See Coral)
    coralline algae, 28
    conclusions, 29
    earbone of cetacean, 29
    echinoid, 28 (See Pyrina)
    ecology of, 29
    gastropods, 28
    rudistids, 26–27 (See Rudistids)
    shark's tooth, 28, 36, 75
    solitary coral, 28
    stromatoporoids, 27–28 (See Stromatoporoids)

  microfossils
    age, affinities, ecology, 30–31 (See Foraminifera)
    faunal lists, 68–70
    Repository of types, 57
Palmer, R. H., 26, 65
Paquier, V., 26, 65
Park, C. F., Jr., 37
Parona, C. F., 27
Patterson, C. C., 35
Pelecypods (See also rudistids, Paleontology—megafossils, Caprina), 66
Pettersson, H., 81
Phleger, F. B, 81
Phleger Bottom Sampler, 10, 12, 15, 76, 81
Phosphatized limestone, 9, 18, 20, 21, 35, 36, 54, 61, 75, 76, 77, 80
Physiography of the Mid-Pacific Mountains, 4–5
Pilsbry, H. H., 51
Piston-type corer (See Kullenberg corer)
Praecaprina, 26
Pyrina, 18, 22, 28, 77
  keenae, 67, 75

Raitt, R. W., 45, 48, 50
Range chart of fossil fauna, 21
Ranger Bank, 52
Red clay, 12, 14, 76–79
Reef coral (See Coral)
Renard, A. F., 29
Revelle, R. R., v, 45, 57
Rex, R. W., 35
Richter, C. F., 47
Roy, C. J., 35
Rubey, W. W., 35, 44, 45
Rudistids, 22, 26, 27, 33, 51
  affinities of, 26
  age of, 21, 26
  ecology of, 27
  general discussion, 26–27
  systematic descriptions, 65, 66

Sampling equipment and techniques, summary, 81
Sampson, E., 42
Sandstone, 11, 20, 76–80
Schenck, H. G., 24
Schuchert, C., 47
Scripps Institution of Oceanography, v, vi, 12, 35, 81
Seamounts (See Guyots)
Shark's tooth (Oxyrhina ?), 28, 36, 75
Shepard, F. P., v, 33, 39, 42, 48, 81, 83
Shimer, H. W., 27, 63

INDEX

Shrock, R. R., 27, 63
Side slopes of seamounts and guyots, 2, 5, 9, 12, 14, 20, 39
   corrections to, 83, 84
Silicified limestone, 9, 35, 37, 38, 54, 75, 76
Simpson, G. G., 50
Skottsberg, C., 51
Snapper sampling device, 81
Stearns, H. T., 42, 43, 50
Steiner, A., 27
Stromatopora, 22, 27
   sp., 62, 73
Stromatoporoids, 18, 22, 27
   age of, 27
   ecology of, 27
   systematic descriptions, 61–63
   taxonomic position of, 27
Submergence of guyots
   causes of, 44–48
   date of, 26
   Gulf of Alaska, 3
   Northern Marshalls, 3–4
   speed of, 25, 26
Subsidence (*See* Submergence)
Suess, E., 23
Sugiyama, T., 27, 62, 63
Sunken Continents in the Mid-Pacific, 51
Sverdrup, H. U., 24, 52
Swinnerton, H. H., 27
Sylvania Seamount, 34, 49

Taliaferro, N. L., 35
Tethyan Province, 23, 29, 30, 31
Tethys Sea, 23, 26, 29, 52
Tracey, J. I., Jr., 33, 48
Trochus, 22, 28
   sp., 65, 71

Truncation of guyots, 41–43
   age of, 31, 41–43
   erosional debris as proof of, 32–33
   history of, hypotheses as to, 42, 43
   morphology and coral as proof of, 41
Turbidity currents, 30
TUSCARORA, USS, 4

Umbgrove, J. H. F., 33, 44, 45
U. S. Coast and Geodetic Survey, 4
U. S. Navy Electronics Laboratory, v, 1, 57, 81
U. S. Navy Hydrographic Office, 83
   sounding sheets of, v, 2, 4
Urgonian, 23, 24
USS EPCE(R) 857, v
USS CAPE JOHNSON, 14
USS TUSCARORA, 4

Vanderhoof, V. L., 29
Vaughan, T. W., 23, 24, 48, 51, 52, 58, 59, 60, 61
Vening Meinesz, F. A., 46, 47, 48, 54
Vermicularia, 18, 22
   sp., 63, 74, 75, 77
Volcanoes, guyots as (*See* Geomorphology)

Wallace, A. R., 50
Washington, H. S., 32
Wegener, A. L., 50
Wells, J. W., 23, 24, 27, 51, 52, 57, 58, 59, 60, 61
Wharton, A. J. L., 41, 49
Williams, H., 40
Woods, H., 27

Yabe, H., 27, 62, 63